ENERGIES

Also by Vaclav Smil

China's Energy
Energy in the Developing World
Energy Analysis in Agriculture
Biomass Energies
The Bad Earth
Carbon Nitrogen Sulfur
Energy Food Environment
Energy in China's Modernization
General Energetics
China's Environmental Crisis
Global Ecology
Energy in World History
Cycles of Life

VACLAV SMIL

ENERGIES

An Illustrated Guide to
the Biosphere and Civilization

The MIT Press
Cambridge, Massachusetts
London, England

This book was set in Galliard by Graphic Composition,
Inc. and printed and bound in the United States of
America.

Library of Congress Cataloging-in-Publication Data

Smil, Vaclav.
    Energies: an illustrated guide to the biosphere and
    civilization / Vaclav Smil.
        p.   cm.
    Includes bibliographical references and index.
    ISBN 0-262-19410-4 (hardcover: alk. paper)
    1. Power resources—Social aspects. I. Title.
TJ163.2.S618 1998
531'.6—dc21   98-13408
                                                        CIP

To Philip and Phylis Morrison

Contents

ACKNOWLEDGMENTS

Habeunt sua fata libri. . . . This one almost did not get written. I wrote its first proposal in 1987 but as there were no immediate takers I set it aside and turned to other energy, and non-energy, projects. A few years ago — sitting in the old MIT Press offices and outlining another of my long-standing book ideas — I mentioned to Larry Cohen, as an afterthought, that I would also eventually like to do an unorthodox book on energies. I described rapidly its contents — and Larry chose it without hesitation.

Marjorie Halmarson, whose graphic art has illustrated many of my books, had a particularly challenging task this time, as she had to prepare more than 300 images. Deborah Cantor-Adams took the book through the editorial process, and Jean Wilcox gave it its good looks. And by dedicating it to Philip and Phylis Morrison, I am repaying a small part of a great personal and intellectual debt.

ENERGY'S UNIQUE ALLURE:
AN UNORTHODOX LOOK

Energy is the only universal currency: one of its many forms must be transformed to another in order for stars to shine, planets to rotate, plants to grow, and civilizations to evolve. Recognition of this universality was one of the great achievements of nineteenth-century science, but, surprisingly, this recognition has not led to comprehensive, systematic studies that view our world through the powerful prism of energy.

Modern energy studies, mirroring the increasing fragmentation of scientific understanding, are splintered into subdisciplines whose practitioners do not pay much attention to one another's writings. Geologists trying to understand the grand surface-forming forces of inexorably moving tectonic plates are rarely aware of the findings of modern bioenergetics, whose reach now extends from studies of elite cyclists to the hovering of hummingbirds; engineers preoccupied with the efficiencies of electricity-generating plants think little about fundamental energy constants and changes that determined the evolution of societies before the emergence of fossil-fueled civilization.

This book attempts to bridge these gaps. Its basic idea is to offer a comprehensive and integrated survey of the energies shaping our world, from the Sun to pregnancy, from bread to microchips. Naturally, such a sweep demands both a logical progression and selectivity.

The first requirement is met by following an evolutionary sequence progressing from planetary energy flows to the lives of plants and animals, then to human energetics and energy in the development of preindustrial and modern societies, and concluding with intensive transportation and information flows, the two most distinguishing characteristics of fossil-fueled civilization.

The chosen topics have been distilled into mini-essays related through frequent cross-referencing (indicated by boldface italics) to other key energy flows and stores. The conflicting needs of technical detail and readability are reconciled through the use of vignettes to cover topics of special interest.

Each entry contains quantitative information expressing essential physical, chemical, or biological characteristics as well as unusual or surprising facts, such as the incredibly high metabolic intensity of bacteria, the astonishingly low energy cost of pregnancy in traditional societies, James Watt's role in *obstructing* the development of steam engines, and the nearly Sun-like power density with which the most powerful microchips discard heat.

I have tried to use images to construct a parallel text, highlighting the most important aspects of energy conversions while also offering a visual and historical context for the discussion. In addition, graphs of long-term trends summarize changes across decades and centuries.

The book is thus a hybrid combining a quasi-encyclopedic sweep with the brevity of mini-essays, technical discourse with accessible descriptions enlivened by original illustrations and reproductions of archival art and modern images. Its origins are in a scientist's fascination with the complexity of the biosphere and the intricacies of human energetics, with technical progress across millennia, and with the achievements of our fossil-fueled civilization. Its aim will be accomplished if readers feel at least some of this fascination.

An Overview of
Concepts and Units

A deep understanding of the peculiarities and complexities of different forms of energy and their stores and conversions requires quantification of these qualities and processes. For this, we must introduce a certain number of scientific concepts and measures and their associated units.

The first problem we encounter in formulating a way to talk about energies is that the common usages of many of the key terms are misleading. As Henk Tennekes has noted, "We have made a terrible mess of simple physical concepts in ordinary life." Few of these muddles are as ubiquitous and annoying as those involving terms such as "energy," "power," and "force."

A knowledge of basic mechanics helps get us started in sorting these terms out. *Force* is defined as the intensity with which we try to displace—push, pull, lift, kick, throw—an object. We can exert a large force even if the huge boulder we are trying to push remains immobile. We accomplish *work*, however, only when the object of our attention moves in the direction of the applied force. In fact, we define the amount of work performed as the prod-

uct of the force applied and the distance covered. *Energy*, as the common textbook definition puts it, is "the capacity for doing work" and thus will be measured in the same units as work. If we measure force in units of newtons (N, named for Sir Isaac Newton) and distance in meters (m), our measure of work will be the awkward-sounding Newton-meter. To simplify, scientists call one Newton-meter a joule (J), named for James Prescott Joule (1818–1889), who published the first accurate calculation of the equivalence of work and heat. The joule is the standard scientific unit for energy and work. *Power* is simply a rate of doing work, that is, an energy flow per unit of time; its measure is thus joules per second. We call one joule per second a watt (W) after James Watt (1736–1819), the inventor of the improved steam engine and the man who set the first standard unit of power, which as it happens was not the watt but the horsepower (hp), a unit equal to roughly 750 W.

To go further we need to move from pushing and shoving (which we call mechanical or kinetic energy) to heating (thermal energy). We define a unit called the calorie as the amount of heat needed to raise the temperature of one gram of water from 14.5 to 15.5°C. (You needn't worry about why we define it this way.) Using this unit will help us compare thermal energies, but again it does not offer an all-encompassing measure that will allow us to compare different forms of energy.

At this point you might be asking, What is energy? This turns out to be a hard question to answer. Even one of the grand summations of modern physics is of little help: "It is important to realize that in physics today, we have no knowledge of what energy *is*. We do not have a picture that energy comes in little blobs of a definite amount," wrote Richard Feynman in his famous *Lectures on Physics*. If forced to choose, I would opt for David Rose's evasive answer: Energy "is an abstract concept invented by physical scientists in the nineteenth century to describe quantitatively a wide variety of natural phenomena."

Our modern understanding of energy includes a number of profound realizations: that mass and energy are equivalent; that many conversions link various kinds of energies; that no energy is lost during these conversions (this is the first law of thermodynamics); and that this conservation of energy is inexorably accompanied by a loss of utility (the second law of thermodynamics). The first realization—initially called an "amusing and attractive thought" in a letter Einstein wrote to a friend—is summed up in perhaps the best known of all physical equations: $E = mc^2$.

The second realization is demonstrated constantly by myriads of energy conversions throughout the universe. Gravitational energy sets galaxies in motion, keeps the Earth orbiting around the Sun, and holds down the atmosphere that makes our planet habitable. Conversion of nuclear energy within the Sun releases an incessant stream of electromagnetic (solar, radiant) energy. A small share of that energy reaches the Earth, which itself also releases

geothermal energy. Heat from both these processes sets in motion the atmosphere, the oceans, and the Earth's huge tectonic plates.

A tiny share of the Sun's radiant energy is transformed through photosynthesis into stores of chemical energy that are used by many kinds of bacteria and by plants. Heterotrophs—organisms ranging from bacteria, protists, and fungi to mammals—ingest and reorganize plant tissues into new chemical bonds and use them also to generate mechanical (kinetic) energy. Chemical energy stored over many millions of years in fossil fuels is released through combustion in boilers and engines as thermal energy, which many processes then convert into mechanical, chemical, or electromagnetic energy.

The second law of thermodynamics addresses the inescapable reality that the potential for useful work steadily diminishes as we move along energy conversion chains. There is a measure associated with this loss of useful energy, and it is called *entropy*: while energy is conserved in any conversion, the conversion can only increase the entropy of the system as a whole. There is nothing we can do about this decrease of utility. A barrel of crude oil is a low-entropy store of very useful energy that can be converted to heat, electricity, motion, and light. Hot air molecules leaving an engine exhaust or surrounding a light bulb represent a high-entropy state in which there is an irretrievable loss of utility.

Loss of complexity and the rise of homogeneity are the unavoidable consequences of this unidirectional entropic dissipation in any closed system. (You can see this if you compare the multitude of complex organic molecules making up crude oil with the sameness of the few kinds of simple molecules making up hot exhaust gas.) In contrast, all living organisms—from bacteria to civilizations—are open systems, constantly importing and exporting energy, and hence are able to maintain themselves in a state of chemical and thermodynamic disequilibrium. They are temporarily defying the entropic trend as their growth and evolution bring greater heterogeneity and higher complexity.

Using unadjusted units to quantify this multitude of processes would be inconvenient: actual figures would nearly always be either trailed or preceded by many zeros. Both joules and watts represent very tiny amounts of measured energy and power: about thirty micrograms of coal—or two seconds' worth of a vole's metabolism account for one joule. One watt is the power of a very small burning candle or a hummingbird's rapid flight.

Multiples are inevitable, and we therefore introduce a series of prefixes to abbreviate the most useful multiples: a kilogram of good coal contains nearly thirty million joules, or thirty megajoules (MJ), of energy, and the world now consumes fossil fuels at the rate of roughly ten trillion watts (TW). We attach the same prefixes to the units that we use to measure electrical energy: Volts (V), which are a measure of the difference in electric potential between two points of a conductor, and amperes (A), which measure the intensity of an electric current. The power of an electrical system is the product of voltage and current, which means that one volt-ampere is also one watt.

Table 1 lists the complete set of multiples as well as the submultiples, which are needed much less frequently when we are dealing with everyday energy flows. Magnitudes of some energy stores and flows are listed in tables 2 and 3. Examples of power ratings of continuous energy conversions are given in table 4, and those of ephemeral phenomena are shown in table 5.

Rates and ratios are important for understanding energy stores, flows, and effects. *Energy density*—the amount of energy stored in a unit mass of a resource (joules per kilogram, J/kg)—is a useful comparative measure for foodstuffs and fuels. Humans would need a daily intake of many kilograms of low-energy-density foods

Table 1 Prefixes of Scientific Units

Prefix	Abbreviation	Scientific notation
deka	da	10^1
hecto	h	10^2
kilo	k	10^3
mega	M	10^6
giga	G	10^9
tera	T	10^{12}
peta	P	10^{15}
exa	E	10^{18}
deci	d	10^{-1}
centi	c	10^{-2}
milli	m	10^{-3}
micro	μ	10^{-6}
nano	n	10^{-9}
pico	p	10^{-12}
femto	f	10^{-15}
atto	a	10^{-18}

Table 2 Energy Stores

Energy of	Magnitude
Global coal resources	200,000 EJ
Global plant mass	10,000 EJ
Latent heat of a thunderstorm	5 PJ
Coal load in a 100-t hopper car	2 TJ
Barrel of crude oil	6 GJ
Bottle of white table wine	3 MJ
A small chickpea	5 kJ
Fly on a kitchen table	9 mJ
A 2-mm raindrop on a blade of grass	4 μJ

Table 3 Energy Flows

Energy of	Magnitude
Solar radiation reaching the Earth	5500000 EJ
Global net photosynthesis	2000 EJ
Global fossil fuel production	300 EJ
Typical Caribbean hurricane	38 EJ
Largest H-bomb tested in 1961	240 PJ
Latent heat of a thunderstorm	5 PJ
Hiroshima bomb of 1945	84 TJ
Basal metabolism of a large horse	100 MJ
Daily adult food intake	10 MJ
Striking a typewriter key	20 mJ
Flea hop	100 nJ

Table 4 Powers of Continuous Phenomena

Energy flows	Power
Global intercept of solar radiation	170 PW
Wind-generated waves on the ocean	90 PW
Global gross primary productivity	100 TW
Global Earth heat flow	42 TW
Worldwide fossil fuel combustion	10 TW
Florida Current between Miami and Bimini	20 GW
Large thermal power plant	5GW
Basal metabolism of a 70-kg man	80 W

Table 5 Powers of Short-Lived Phenomena

Energy flows	Duration	Power
Richter magnitude 8 earthquake	30 s	1.6 PW
Large volcanic eruption	10 h	100 TW
Thunderstorm's kinetic energy	20 min	100 GW
Large WWII bombing raid	1 h	20 GW
Average U.S. tornado	3 min	1.7 GW
Four engines of Boeing 747	10 h	60 MW
Watt's largest steam engine	10 h	100 kW
Running 100-m dash	10 s	1.3 kW
Machine-washing laundry	20 min	500 W
Playing a CD	60 min	25 W
Candle burning to the end	2 h	5 W
Hummingbird flight	3 min	0.7 W

such as fruits and vegetables to maintain themselves, whereas less than half a kilogram of rice, which has a high energy density, will do. Conversely, gasoline makes a great portable fuel because its energy density is nearly three times that of air-dried wood. Table 6 shows the energy densities of some common fuels, foods, and their metabolic products.

Power density—the rate at which energies are produced or consumed per unit of area (watts per square meter, W/m^2)—is a critical structural determinant of energy production systems. The power density of fuel production from a large open-cast mine extracting excellent bituminous coal from a thick seam is easily more than 1000 W/m^2; the power density of electricity generation in a large hydrostation whose dam creates the huge reservoir needed to store a sufficient volume of water may be as low as a few W/m^2. In order to illustrate the spatial aspect of various energy conversions, values of power densities are

Table 6 Ranges of Energy Densities of Common Fuels and Foodstuffs

Energy density	(MJ/kg)
Hydrogen	114.0
Gasolines	46.0–47.0
Crude oils	42.0–44.0
Pure plant oils	38.0–37.0
Natural gases	33.0–37.0
Butter	29.0–30.0
Ethanol	29.6
Best bituminous coals	27.0–29.0
Pure protein	23.0
Common steam coals	22.0–24.0
Good lignites	18.0–20.0
Pure carbohydrates	17.0
Cereal grains	15.2–15.4
Air-dried wood	14.0–15.0
Cereal straws	12.0–15.0
Lean meats	5.0–10.0
Fish	2.9–9.3
Potatoes	3.2–4.8
Fruits	1.5–4.0
Human feces	1.8–3.0
Vegetables	0.6–1.8
Urine	0.1–0.2

Table 7 Efficiencies of Common Energy Conversions (percent)

Conversions	Energies	Efficiencies
Large electricity generators	M → e	98–99
Large power-plant boilers	c → t	90–98
Large electric motors	e → m	90–97
Best home natural-gas furnaces	c → t	90–96
Dry-cell batteries	c → e	85–95
Human lactation	c → c	85–95
Overshot waterwheels	m → m	60–85
Small electric motors	e → m	60–75
Large steam turbines	t → m	40–45
Improved wood stoves	c → t	25–45
Large gas turbines	c → m	35–40
Diesel engines	c → m	30–35
Mammalian postnatal growth	c → c	30–35
Best photovoltaic cells	r → e	20–30
Best large steam engines	c → m	20–25
Internal combustion engines	c → m	15–25
High-pressure sodium lamps	e → r	15–20
Mammalian muscles	c → m	15–20
Traditional stoves	c → t	10–15
Fluorescent lights	e → r	10–12
Steam locomotives	c → m	3–6
Peak crop photosynthesis	r → c	4–5
Incandescent light bulbs	e → r	2–5
Paraffin candles	c → r	1–2
Most productive ecosystems	r → c	1–2
Global photosynthetic mean	r → c	0.3

Energy labels: c—chemical, e—electrical, m—mechanical (kinetic), r—radiant (electromagnetic, solar), t—thermal

shown in figures 1 and 2, graphs plotting areas against power.

The *efficiency* of an energy conversion is the ratio of the amount of energy output to the amount input. This measure is used to describe the performance of energy converters such as boilers, engines, photovoltaic cells, or lights. Efficiencies of common converters are listed in table 7. **Energy intensity** is the cost of a product or service in energy terms. Titanium and aluminum are highly energy-intensive, for example, whereas iron and glass are fairly cheap. Typical ranges of the energy costs of common materials are given in table 8.

Technical advances keep pushing efficiencies up and energy intensities down: today's best lights are about twenty times as efficient as the first light bulbs of more than a hundred years ago, and production of a kilogram of steel now requires less than one-tenth the energy it did two centuries ago.

The best way to understand the world of energy and power units is to get a small calculator and do exercises rooted in the realities all around us. How much energy does it take to play Mozart's last piano concerto? What volume of gasoline contains energy equivalent to a cord of stacked, air-dry wood? What is the kinetic power of an arrow shot from a modern compound bow? How much more powerful are the four engines of a Boeing 747 than the eight engines on B-52 bomber? What share of daily metabolism can be supplied by a slice of whole-wheat bread?

If you are so inclined, the figures listed in the tables given here, the ranges shown in the power density figure, and the hundreds of numbers scattered through the book will lead you to a virtually unlimited supply of such challenges—and hence to an insider's understanding of energies.

Table 8 Typical Energy Costs of Common Materials (MJ/kg)

Material	Energy cost	Made or extracted from
Aluminum	227–342	Bauxite
Bricks	2–5	Clay
Cement	5–9	Clay and limestone
Copper	60–125	Sulfide ore
Glass	18–35	Sand, etc.
Iron	20–25	Iron ore
Limestone	0.07–0.1	Sedimentary rock
Nickel	230–70	Ore concentrate
Paper	25–50	Standing timber
Polyethylene	87–115	Crude oil
Polystyrene	62–108	Crude oil
Polyvinylchloride	85–107	Crude oil
Sand	0.08–0.1	Riverbed
Silicon	230–235	Silica
Steel	20–50	Iron
Sulfuric acid	2–3	Sulfur
Titanium	900–940	Ore concentrate
Water	0.001–0.01	Streams, reservoirs
Wood	3–7	Standing timber

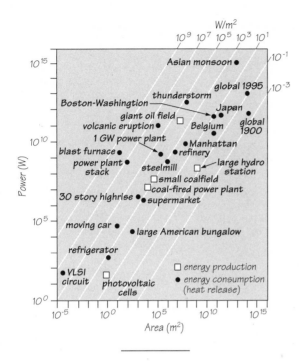

Figure 1
Power densities of various energy production
and consumption phenomena.

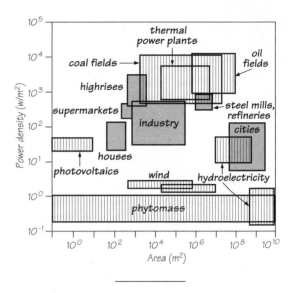

Figure 2
Typical ranges of areas and power densities in large-scale
modern energy production (lines) and in household
and industrial consumption (dots).

1

SUN AND EARTH

As long as the *Sun* has enough hydrogen in its core to sustain orderly thermonuclear reactions, the star will flood the *Earth* with a surfeit of *solar radiation* that will continue to energize most of the physical and chemical processes on our planet. This *radiation* heats the *atmosphere* and the *ocean*, generates *winds* and *rains*, and powers the inexorable advances of surface *denudation*. Of all the Earth's global energy transformations, only *geotectonics*—a slow reworking of the planet's *ocean* floor and continents accompanied by *earthquakes* and by spectacular energy releases in *volcanoes*—is driven not by the stellar *radiation* but by gravity and by the gradual release of the *Earth*'s *heat*.

Sunlight also energizes *photosynthesis*, the planet's most important biochemical conversion, creating new biomass in *bacteria, phytoplankton,* and higher *plants,* and above all in *forests* and *grasslands.* This synthesis is the foundation of *food chains* providing the nutrition needed for the *heterotrophic metabolism* of *animals* and *people,* nutrition that allows activities ranging from simple *running* to elaborate tasks of *labor and leisure.*

Human societies—from small groups of *hunters and gatherers* to complex societies dependent on enormous flows of *fossil fuels and electricity*—have all been inextricably tied to the steady flux, and converted stores, of sunlight. Accumulated stores of photosynthetically converted solar radiation, tapped as *coals, crude oils,* and *natural gases,* will continue to serve for many generations as the foundation of *fossil-fueled civilization* with its profusion of energy services, ubiquitous *transportation,* and surfeit of *information.*

Sun

An extraterrestrial observer could find nothing to distinguish the Sun among the millions of similar stars that are in turn only a small fraction of the hundred billion radiant bodies of our galaxy. Our star belongs to a common class of stars located roughly in the middle of the main sequence of the Hertzsprung-Russell classification scheme: it is a "G2 dwarf" of characteristic yellow color and of unremarkable stellar magnitude (+4.83).

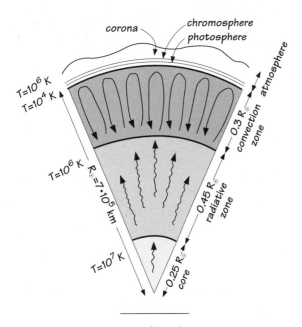

Cross-section of the solar interior.

The earliest scientific explanation of *solar radiation*—Hermann Helmholtz's calculation based on gravitation—yielded about thirty million years for the age of the star. Albert Einstein's famous equation linking mass to energy opened the path toward a correct explanation, but it alone did not give a satisfactory solution. It also seemed unlikely that the complete annihilation of solar matter, transforming nuclei and electrons into *radiation* (an explanation championed by Sir Arthur Eddington), would occur even in temperatures surpassing ten million degrees on the Kelvin scale (K). The idea we accept today—that

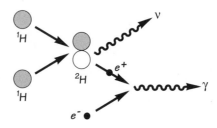

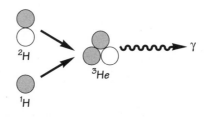

After about 4.5 billion years of life the *Sun* is almost halfway through its transformation from a dwarf to a red giant. Its luminosity will eventually become about a thousand times higher than today, and its vastly expanded diameter will reach the Earth. For a time, the planet will orbit within the star's low-density envelope, but eventually and inevitably it will spiral inward and be swallowed by the red giant's core.

Life on *Earth* will be eliminated well before the *Sun* turns into a red giant: as its core contracts and thermonuclear reactions heat its outer layer, the star's diameter will expand tenfold, and *radiation* from the red subgiant will vaporize all *oceans* on the *Earth* and will generate fierce hot *winds* in the planet's convulsed *atmosphere.* But as long as there is hydrogen at the star's core, the inexorable changes in luminosity will be gradual, and the *Sun* will continue to energize the *Earth*'s life as well as most of its physical transformations.

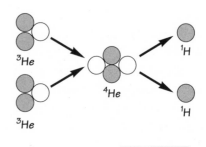

The proton-proton reaction forming ^{4}He and releasing energy.

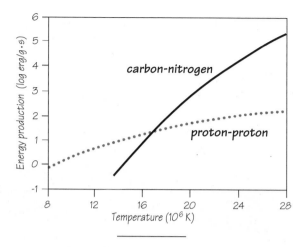

Energy production of the Sun as a function of core temperature.

total solar luminosity (3.89×10^{26} W) and the area of a sphere with the orbital radius (which is, on average, 150 million km).

This flux, traditionally known as the solar constant, is the maximum rate of energy input available at the top of the *Earth*'s *atmosphere*. During the early 1970s NASA used 1353 W/m² as the design value for its space vehicles.

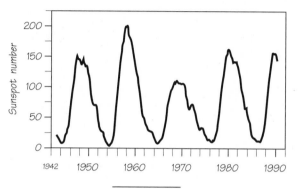

Sunspot numbers, 1947–1990 (average monthly values smoothed with a thirteen-month running mean).

nuclear reactions are involved in the production of energy in the *Sun*'s core—was proposed in the late 1930s by Hans Bethe, Charles Critchfield, and Carl Friedrich von Weizsäcker.

The fusion of hydrogen into helium in the proton–proton cycle begins when the temperature reaches thirteen million K. Just above sixteen million K the carbon-nitrogen cycle, which regenerates ¹²C, starts to dominate. We cannot be sure, but according to the best current models the C-N cycle generates only about 1.5 percent of the Sun's total energy.

Core reactions consume 4.3 million tonnes of matter every second and, according to Einstein's mass-energy relationship, liberate 3.89×10^{26} J of nuclear energy. This immense nuclear flux is rapidly transformed into thermal energy and is transported outward first by random reradiation, then by faster directional convection. Assuming isotropic *radiation,* the power passing through every square meter of the visible light-emitting layer of the photosphere is about 64 MW. There is virtually no attenuation as *solar radiation* streams through space. When it reaches the *Earth*'s orbit, its power density is the quotient of

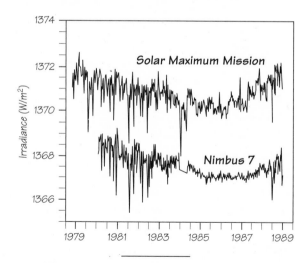

Short-term fluctuations of solar output measured by radiometers on two satellites. On the average, the highest output coincides with the maxima of sunspot activity.

The flux has been measured directly in space since 1979, when the Nimbus 7 satellite recorded values around 1371 W/m². In 1980 the weighted mean from the newly launched Solar Maximum Mission satellite was 1368.3 W/m².

Continuing monitoring from space has revealed a complex pattern of tiny short-term fluctuations impossible to discern from the observations carried through the intervening, and interfering, *atmosphere.* Short-term (on the order of days to weeks) variations of up to 0.2 percent are caused by the passage of dark sunspots and bright faculae that determine the local emissions of the rotating *Sun.* Longer-term results indicate that the star's irradiance declined by about 0.1 percent between the peak and the trough of the eleven-year solar activity cycle.

This seems to be a shift too small to affect the *atmosphere*—and yet there are some very persuasive correlations between the cyclical changes of *solar radiation* and atmospheric oscillations. So far the most plausible explanation of this effect is that it is due to the altered heating of the stratosphere by the ultraviolet *radiation,* whose cyclical changes are more pronounced than the shifts in visible part of solar spectrum.

Questions about the relationship between solar variability and the *Earth*'s climate on time scales from years to centuries will not be settled soon. No less intriguing is the fact that today's solar luminosity is about 40 percent higher than it was 4.5 billion years ago—but that the *Earth*'s climate has remained remarkably constant during nearly four billion years of biospheric evolution.

Solar Radiation

Wavelengths of the *Sun*'s electromagnetic energy streaming toward the *Earth* span more than ten orders of magnitude, from the shortest gamma rays and x-rays of less than 10^{-10} m to radio waves longer than 1 m. The pattern of the solar spectrum resembles closely the *radiation* of a

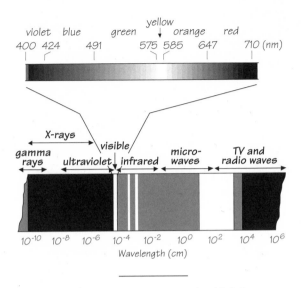

The electromagnetic spectrum and visible light.

perfect black body at 6000 K. This similarity is especially close for wavelengths longer than those of yellow light; for shorter wavelengths the solar spectrum falls noticeably below the 6000 K line. According to Wien's displacement law the maximum emission for this temperature is at 483 nm, close to the end of the blue part of the visible spectrum, near green.

Energy flux is divided unevenly among the three large spectral categories. Ultraviolet (UV) *radiation,* from the shortest wavelengths to 400 nm, accounts for less than 9 percent of the total; visible light, ranging from 400 nm of the deepest violet to 700 nm of the darkest red, for 39 percent; and infrared (IR) *radiation* for about 52 percent. *Radiation* reaching the *Earth*'s surface is very different from the extraterrestrial flux, in both quantity and quality. Key physical realities responsible for these alterations include the elliptical orbit of the *Earth,* its shape and the inclination of its rotation axis, the composition of its *atmosphere,* and the reflectivity (albedo) of clouds and surfaces.

Changes in incoming *solar radiation* caused by the elliptical orbit are negligible in comparison with the consequences of our planet's sphericity, albedo, and rotation. The power density of an otherwise unchanged *radiation* received on the *Earth*'s spherical surface must be only a quarter of the solar constant value, that is, no more than about 342 W/m². Reflection of *radiation* by cloud tops, *oceans,* and terrestrial surfaces varies greatly with seasons, but the average planetary albedo reduces this flux by about 30 percent, to 235 W/m². This would be the annual mean of the *solar radiation* reaching the ground in subtropical desert areas with clean air and minimal cloudiness.

Axial tilt causes predictably regular latitudinal differences in the amount of received radiation. The equatorial belt has only small seasonal variations, whereas the polar

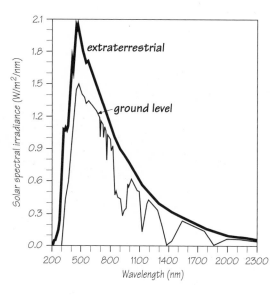

Extraterrestrial and ground-level spectra of solar radiation.

regions oscillate between complete absence and the day-long presence of sunlight. The changing angle of incidence carries the *solar radiation* through different thicknesses of the *atmosphere,* where it is scattered and also absorbed by gases and aerosols.

Consequently, the *solar radiation* reaching the *Earth*'s surface displays complex spatial and temporal patterns. Its annual global mean is just below 170 W/m² for *oceans,* and about 180 W/m² for continents. The most notable departure from expected latitudinal regularity is the solar impoverishment of the tropics, and the monsoonal subtropics, caused by high cloudiness. Large parts of Brazil, Nigeria, and South China get less insolation per year than New England or parts of Western Europe!

Even more surprisingly, there is no difference in the peak midday summer fluxes between equatorial Jakarta and such subarctic cities as Edmonton in Canada or Yakutsk in Siberia. Local cloudiness can also result in large differences over short distances. Perhaps the best example

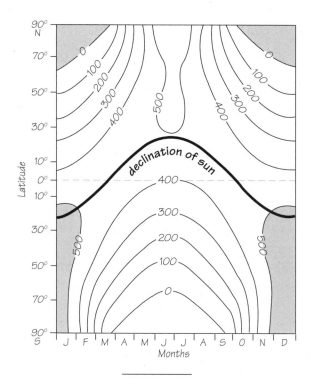

Latitudinal distribution of solar radiation (in W/m²) received at the top of the atmosphere.

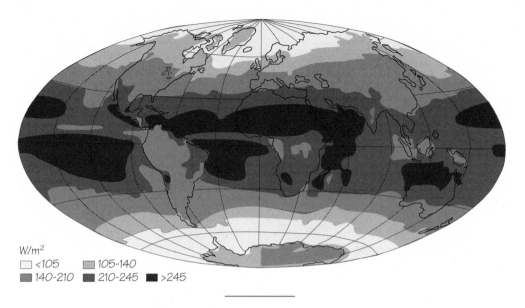

Annual average of total radiation (in W/m²) received at the
Earth's surface, based on eight years of satellite observations.

comes from Oahu, where the cloud-shrouded Koolau Range, which intercepts clouds and *rains* driven by strong trade *winds,* has an annual mean of 150 W/m², whereas Pearl Harbor, just 15 km downwind, averages 250 W/m².

The *solar radiation* average of about 170 W/m² delivers annually a total solar energy input of 2.7×10^{24} J, or roughly 87 PW. This total is almost 8000 times higher than the worldwide consumption of *fossil fuels* and *electricity* during the early 1990s. Only a tiny fraction of this immense flow is absorbed by pigments energizing *photosynthesis,* and a larger, but still very small, part warms the bodies of *plants and animals* and *people* and their shelters. But the *radiation* heating the *ocean* and barren rocks and soils also sustains life by driving critical biospheric services: energizing the water cycle, generating *winds,* maintaining temperature ranges suitable for efficient

functioning of metabolic processes and organic decomposition, and powering weathering and *denudation,* which mobilize mineral nutrients needed for primary production.

Absorbed *solar radiation* must return to space to keep the planet in long-term thermal equilibrium, but it leaves the *Earth* drastically shifted into the infrared range. In contrast to the *Sun*'s shortwave outpouring—determined by its photospheric temperature of 5800 K—the terrestrial *radiation* corresponds closely to the electromagnetic emissions of a black body at 300 K (27°C). This warm sphere radiates with peak irradiance at 9.66 μm, deep in the IR zone. And because about 99 percent of solar irradiance comes in wavelengths shorter than 4 μm while the terrestrial spectrum barely extends below 3 μm, there is only a minimal frequency overlap between these two great streams of energy.

Atmosphere

The *Earth*'s *atmosphere* (troposphere and stratosphere) is so thin that in a drawing of the planet with a 10 cm diameter it would be about 0.4 millimeters thick, merely a thin pencil line. Yet this thin gaseous envelope is critical for establishing the energy balance of the *Earth:* the planet is fit for life because its *atmosphere* is so strikingly different from those of its two closest neighbors. The *atmosphere* of Venus is 96 percent CO_2, with 3.5 percent nitrogen and traces of noble gases, water, and ozone. Mars has 95.3 percent CO_2, 2.7 percent nitrogen, 1.6 percent argon, and also traces of water and O_3. A similar terrestrial *atmosphere* would result in average ground temperatures in excess of 200°C and a surface pressure of a few MPa. Such conditions would preclude the existence of complex, carbon-based life assembled in moist tissues.

There is little doubt that the *Earth*'s early *atmosphere* was CO_2-rich, but it is not clear whether its subsequent removal was due exclusively to inorganic geochemical processes (above all carbonic acid runoff) or whether early organisms were important in sequestering CO_2 into $CaCO_3$ sediments. In contrast, there is little doubt that

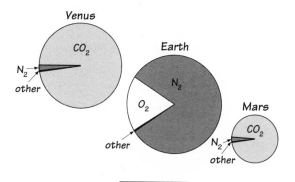

Uniqueness of the Earth's atmosphere in comparison with its two nearest planetary neighbors.

photosynthesis, carried initially by *bacteria,* was responsible for transforming an anoxic Archean *atmosphere.*

An increase of oxygen began accelerating about 2.1 billion years ago, and the present level of 20 percent was reached about three hundred million years ago. Rising levels of tropospheric oxygen also allowed the formation of stratospheric ozone, which shielded the biosphere from high-energy UV *radiation* with wavelengths below 295 nm. Without this protection there would have been no evolution of more complex *plants and animals* because the highest UV frequencies are lethal to most organisms. Lower ones kill germs but also burn skin.

Human actions can do little to upset the shares of major atmospheric constituents. Removal of *nitrogen* for synthesis of ammonia takes out a negligible fraction of the vast tropospheric stores—and denitrification eventually recycles all of the gas. Even a complete exhaustion of all known reserves of *fossil fuels*—an impossible feat owing to prohibitive costs of advancing extraction—would reduce the atmospheric concentration of O_2 by less than two percent. Local and regional air pollution involves many gaseous emissions, but risks of a global

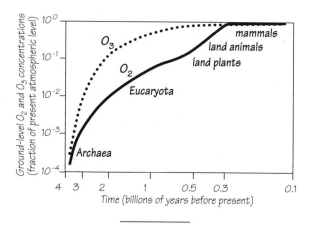

Evolution of atmospheric O_2 and O_3 levels.

climatic change come only with higher releases of trace compounds.

A number of these gases—above all CO_2, N_2O, and CH_4—as well as water vapor, absorb strongly in IR spectrum. Consequently, IR *radiation* leaves the *Earth*'s surface through a number of distinct windows interspaced with absorption bands. Water vapor has the strongest absorption bands, between 2.5 and 3 μm and between 5 and 7 μm, whereas CO_2 has two narrower spikes around 2.5 and 4 μm and a broader band around 15 μm. Because the terrestrial *radiation* is wholly within the IR spectrum, this absorption has a great effect on the *Earth*'s *radiation* balance.

Only very small concentrations of "greenhouse" gases are needed to keep the biosphere hospitable to life. CO_2 now accounts for only about 360 ppm (less than 0.04 percent) of the *Earth*'s *atmosphere*, and other trace gases are measured merely in ppb or ppt. This composition results in mean planetary ground temperature of about 16°C. Combined with surface pressure around 101 kPa, this assures the liquidity of water and temperatures suitable for *photosynthesis* and *heterotrophic metabolism*.

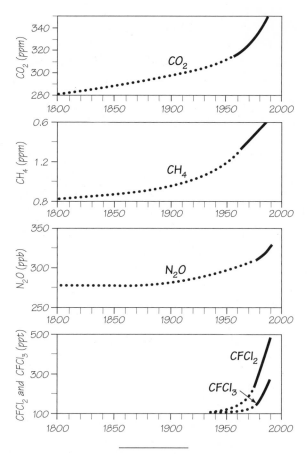

Increasing concentrations of "greenhouse" gases generated by human activities.

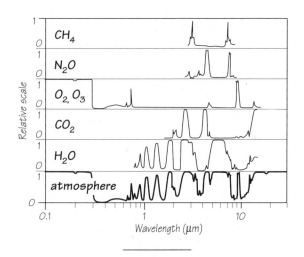

Absorption of terrestrial radiation by atmospheric gases.

Rising concentrations of trace gases can gradually increase average tropospheric temperatures. Conversion of *forests* and *grasslands* to croplands and combustion of *fossil fuels* have been releasing larger amounts of CO_2, while expanding use of *nitrogen* fertilizers, larger cattle herds, and increased rice cultivation release additional N_2O and CH_4. Besides their destructive effects on stratospheric ozone, chlorofluorocarbons are also potent greenhouse gases. The combined effects of anthropogenic greenhouse gases have already reached an average global thermal forcing equivalent to about 2.5 W/m² over large areas of northern hemisphere—but we are not sure how far and how fast this trend will progress.

The *atmosphere* also affects the planet's energy balance by redistributing both sensible and latent heat through *winds* and *rains,* and, in a very different but clearly the most spectacular way, by lightning. Most of these exceedingly concentrated energy discharges originate in cumulonimbus clouds, and their power increases with the fifth power of the cloud size (doubling the cloud increases the lightning power thirty-fold). Single strokes discharge 20–500 MJ, most of it within just 10 μs, producing astounding power of 1–10 GW/m. Visible light accounts for just 0.2–2 percent of the dissipated energy. Most of the discharge goes into heating the surrounding *atmosphere* and producing the acoustic energy of the thunder. Satellite observations indicate a global mean of about a hundred flashes per second.

Planet Earth

Forces acting on the *Earth* as a body in space have profound energetic implications. Gravitation orders and orients, and hampers or assists, all kinetic energy flows. Rotation generates centrifugal and Coriolis forces: the first one flattens the planet at the poles and distends it along the equator; the second deflects the direction of *winds* and currents in the *oceans* (to the right in the northern hemisphere). Rotation also imparts a distinct daily rhythm to the lives of *plants and animals,* and its deceleration, adding 1.5 ms to the average day per century, converts into some three terawatts of tidal friction.

But neither gravitation nor rotation make the *Earth* unique among the celestial bodies. That uniqueness comes from the planet's internal thermal properties, which drive the cycles of surface-altering *geotectonics,* and from its modification and conversion of the incoming *solar radiation* by the *atmosphere, oceans,* and *plants.* The origins of these distinctive processes are unclear. Decay of long-lived radioactive isotopes allows us to fix the Earth's age at

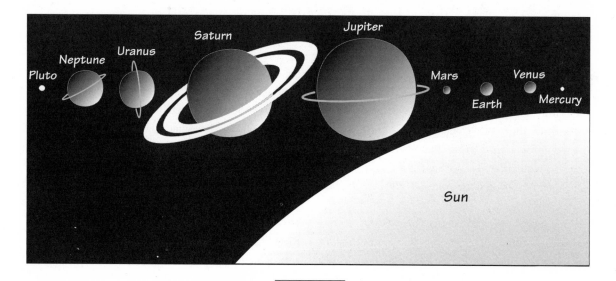

Approximate relative sizes of the planets and the Sun.

Earthrise above the Moon.

4.5 billion years, but hardly anything is certain about the formation of the planet or about the energetics of the early *Earth*. Formation of the *Sun* from a gravitational instability in a dense interstellar cloud and the subsequent agglomeration of planets from the remaining orbiting matter is a highly plausible account of the solar system's origin, but it is not clear if the earliest *Earth* was extremely hot or relatively cool.

Basic geophysical uncertainties extend to the present. Alternative answers to how much ⁴⁰K is in the *Earth*'s core, and how mantle convection works (as a single cell, or in two stages?) provide great differences in understand-

ing and explaining the *Earth*'s *heat* flow and the planet's *geotectonics*. But at less than 100 mW/m², the *Earth*'s internal *heat* flux is too small to affect the planetary energy balance, which is determined by the reflection, absorption, and return of *solar radiation*.

The *radiation* balance of the *Earth* (R_p) at the top of the *atmosphere* is set by the sum of the extraterrestrial irradiance (Q_o, the solar constant) reduced by the planetary albedo and the total outgoing longwave flux (Q_l): $R_p = Q_o(1 - a_p) + Q_l = 0$. The emitted flux is the sum of atmospheric and surface *radiation:* $Q_l = Q_{ea} + Q_{es}$. The *radiation* balances of the *atmosphere* (R_a) and the *Earth*'s surface (R_s) are the differences of their respective absorptions and emissions: $R_a = Q_{aa} + Q_{ea}$ and $R_s = Q_{as} + Q_{es}$ so that $R_p = R_a + R_s = 0$. Global means (all in W/m²) are 342 for Q_o, 107 for a_p, and –235 for Q_l, 67 for Q_{aa}, and 168 for Q_{as}. Outgoing radiation (Q_l, totaling 235 W/m²) is the sum of fluxes reradiated by the *Earth*'s surfaces (40 W/m²) and emitted by the *atmosphere* (195 W/m²).

Net *radiation* flux at the *Earth*'s surface averages about 100 W/m², ranging from midocean peaks surpassing 150 W/m² to zero in the Central Arctic and to slightly negative values (around –5 W/m²) in Antarctica. Most of this energy is lost to the *atmosphere* as latent heat flux driving the global water cycle. But the most important source of atmospheric heating is the reradiation of longwave terrestrial heat. Latent heat flux is a secondary contribution, and sensible heat flux is important only in arid regions where there is not enough water for evaporation.

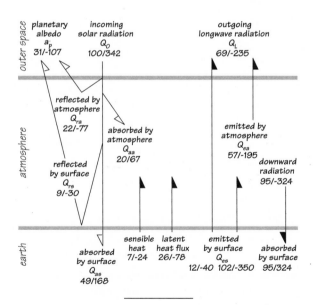

Planetary radiation flows as the shares of total solar radiation and as annual means in W/m².

Oceans and continents also receive most of their heat indirectly as longwave (4–50 μm) emissions reradiated by the *atmosphere.*

This downward flux is emitted by greenhouse gases: fluctuating concentrations of water vapor dominate the emissions with 150–300 W/m², and CO_2 contributes around 75 W/m². The longwave exchange between the surfaces and the *atmosphere* delays only temporarily the outgoing emissions of terrestrial heat, but it controls biospheric temperatures. Its maxima are almost 400 W/m² in cloudy tropics and it remains important in all seasons, although day-to-day variations may be large. Even a single passing cloud can increase the flux by 25 W/m². Higher anthropogenic emissions of greenhouse gases have increased this flux by about 2.5 W/m² since the late nineteenth century.

As expected, satellite observations confirm that the energy balance of the *Earth* (R_p) is in phase with the incoming *solar radiation* (Q_o) — but the global mean of outgoing *radiation* (Q_l) is out of phase with the irradiance, peaking during the northern hemisphere's summer. Asymmetric land-sea distribution explains most of this behavior. Because of its much larger land mass the northern hemisphere experiences much larger seasonal temperature changes and it dominates the global flux of outgoing *radiation.*

Perhaps the most surprising result of satellite monitoring is that over the course of a year the hemispheres are in independent *radiation* balance with the outer space. Of course, seasonally, they have predictable *radiation* deficits and surpluses and during some months the planetary *radiation* balance is not equal to zero for the entire Earth. Also, the atmospheric contribution to the poleward energy transport is asymmetric with the respect to the equator: the extremes are about 3 PW near 45° N and –3 PW near 40° S.

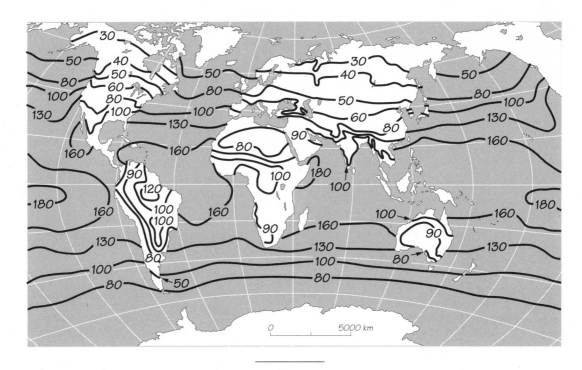

Radiation balance for the Earth's surface in W/m².

Winds

Global *winds* are related to the total *solar radiation* much like the energies of *earthquakes* or *volcanoes* are related to the total flux of the *Earth's heat:* only a tiny part of the overall flow is sufficient to generate spectacular phenomena with immense impacts on the biosphere. Turbulence aloft and friction at the surface dissipate *wind's* kinetic energy and continuous conversion of *solar radiation* to heat and generation of substantial pressure differences are required to maintain the *atmosphere* in motion. No more than 2 percent, and perhaps as little as 0.7 percent, of total solar irradiance is needed to sustain the planet's *winds.*

Their grand global pattern is determined above all by the combination of air flows within the Hadley cell and the deflecting effect of Coriolis force associated with the *Earth's* rotation. Vigorous tropical heating drives the as-

cent of equatorial air in the cell; the air moves poleward in high altitudes, and then returns in the form of persistent trade *winds.* Their sustained average ground speeds are five to six meters per second, and much more at higher altitudes: standing at the crest of Oahu's Koolau range gives you an unforgettable experience of their force. In higher latitudes, including most of the densely inhabited parts of North America and Eurasia, *winds* near the ground are dominated by the westerlies with speeds commonly close to ten meters per second.

Strongest tropospheric *winds* do not arise from these grand circulations of the *atmosphere,* but rather from intensive local or regional heating spawning rapidly moving low-pressure (cyclonic) flows ranging from rainstorms to hurricanes. Localized thunderstorms are the most common, and the least powerful, cyclonic flows. Their downdrafts spread at the surface into gusts of fifteen to twenty-five meters per second, they last for relatively brief periods, and individual cells affect areas no larger than 50–150 square kilometers. Kinetic energy of their most intensive phase is discharged mostly high in the troposphere, and their power will reach 75–600 GW. Gusting *winds* near the ground may repeatedly expose vertical surfaces to power densities of up to 15 kW/m², too low to cause any severe damage.

Severe storms are a different matter: they can last for many hours, cover areas two orders of magnitude larger than typical events, and develop downbursts of up to 70 meters per second. Their *rains* can set new hourly or daily precipitation records for the affected areas. Fortunately, in North America less than three percent of all storms fit into the severe category.

Hurricanes are the most powerful, as well as most extensive, cyclones. Every year about half a dozen of these

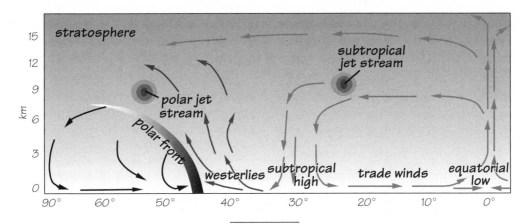

Cross-section of the atmosphere showing its
general circulation pattern.

Hurricane Iris on September 3, 1995.

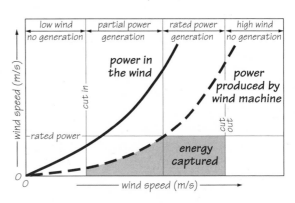

Limits on energy captured by wind machines.

cyclones sweep the Western Atlantic between Venezuela, the Gulf of Mexico and Maine, and about twenty of them affect East Asia. Much rarer Arctic hurricanes can develop under special circumstances in the northernmost Pacific. Tropical hurricanes can wax and wane for days as they traverse land and water their (on the northern hemisphere) counterclockwise swirls can cover up to one million km².

Their *winds* reach sustained speeds of 30–50 and maxima of up to 90 meters per second; on the edge of the cyclonic eye they will expose vertical surfaces to power densities of up to 1 MW/m² which can be withstood only by well-designed modern structures. Although the kinetic energy of *winds* in storms and hurricanes causes damage to vegetation and property that is both severe and extensive, its magnitudes are puny in comparison to the latent heat given off by the *rains* in rapidly advancing cyclones.

Devastating impact of tornadoes is restricted to often very sharply bounded alleys. Studies of many thousands of American tornadoes make it possible to define the average event: its path is about 125 meters wide and nearly 10 kilometers long, its *wind* speed is about 60 meters per second, and it lasts less than three minutes. The power of its 100-meter tall funnel is about 1.7 GW, and *winds* hit vertical surfaces with power densities matching, or surpassing, those of major hurricanes. Rare supertornadoes have *winds* up to 130 meters per second and power of up to 100 GW; their paths, up to two kilometers wide, can persist for more than 150 kilometers.

When judging *wind* as a source of commercial power the enormous total of its global kinetic energy — at about 300 EJ just slightly less than the world's primary energy consumption in 1995 — is an irrelevant value to begin with. A very generous definition of possible global *wind* resource base would include all flows within 1 km of the surface. Their power amounts to about 1.2 PW, and it is unlikely that more than one-tenth of this flux could be actually converted into *electricity* without major atmospheric changes.

Highly uneven distribution of strong *winds* and their large temporal variations have contributed to difficulties in harnessing the flows for *electricity.* Locations with steady *winds* suitable for large-scale commercial conversion are only rarely close to major load centers, and many densely populated regions with rising energy needs have no practical opportunities for such generation.

Rains

Latent heat of *rains* is such an important way of distributing thermal energy through the *atmosphere* because water has an exceptionally high heat of vaporization: 2.43 kJ are needed to evaporate one gram at 20°C. *Solar radiation* can meet this demand in all but the highest latitudes: evaporation of 1 mm/day calls for heat inputs of close to 30 W/m², and annual averages surpass this flux up to about 70° N in the Arctic and to about 60° S in the Antarctic Ocean. Above the *oceans* latent heat flux is limited only by net *radiation,* above the continents it obviously depends also on the supply of water.

Global pattern shows the highest latent heat fluxes (150–180 W/m²) in the Caribbean and along the Gulf Stream in the Atlantic, along the Kuroshio Stream and in two large, separate areas along the equator in the Pacific, and between Madagascar and Australia, in the Bay of Bengal and off the Horn of Africa in the Indian Ocean. Terrestrial maxima above the tropical rain *forests* of South America and Africa rate only about half as much. Global mean of latent heat flux is almost 80 W/m², and hence about 40 PW of *radiation* are used in driving the *Earth*'s water cycle.

This heat input implies worldwide evaporation means of about 3 mm a day, or 1 m a year, with nearly seven-eighths of this total originating from *oceans.* On land *photosynthesis* intensifies evaporation as *plants* speed up the water flow between soils and *atmosphere.* This effect is particularly evident above mature *forests:* in decid-

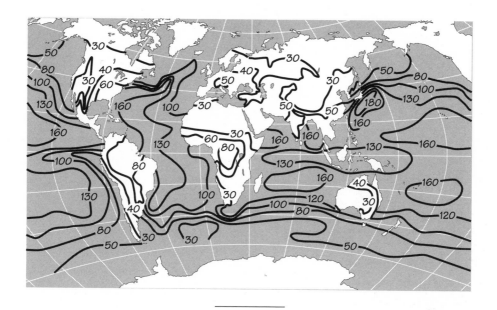

Global patterns of latent heat flux
(solar radiation spent on evaporation in W/m²).

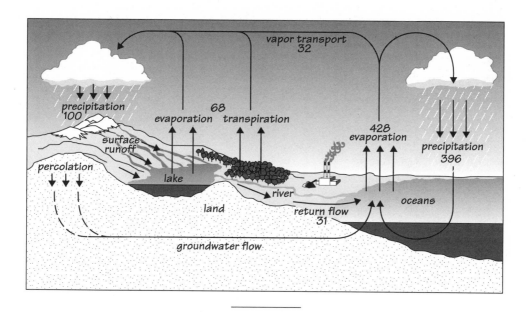

Major flows of the global water cycle.

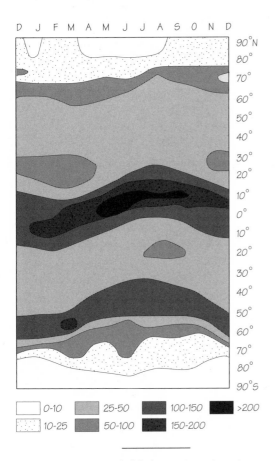

Mean annual global precipitation (in mm).

carry only a few tenths of one percent of the **Earth**'s total latent, a negligible share compared to summer monsoons.

Asian monsoon is the planet's most impressive mechanism for transferring heat absorbed by warm tropical **oceans** to dry land. Nearly half of humanity is directly affected by its seasonal pulse, initiated by intense heating of equatorial waters. The total volume of its **rains** is about 10,000 km³, and they release about five hundred times more latent heat than the most powerful hurricanes. Even so, the Asian monsoon deposits only about one-tenth of all continental **rains,** and less than two percent of total global precipitation.

Roughly three-fifths of all continental precipitation are evaporated, one-tenth returns to **oceans** as surface runoff, and just short of a third is carried by **rivers.** Combined surface and **river** runoff amounts to a bit over 50,000 km³ a year. With the mean continental elevation of about 850 meters, this implies an annual conversion of some 400 EJ of potential energy to kinetic energy of flowing water. This huge flux (nearly 13 TW) — complemented with high

uous woods the latent heat flux may be an order of magnitude higher with peak foliage phytomass than in the late fall.

Latent heat released in cyclonic precipitation far surpasses the kinetic energies of thunderstorms and hurricanes. Even ordinary summer storms precipitating just a few cm of **rain** will release 10–100 times more energy in heating the **atmosphere** than in their **winds.** Latent heat released by hurricanes is typically several thousand times larger than that of summer storms, but a relatively low frequency of these powerful cyclones means that they

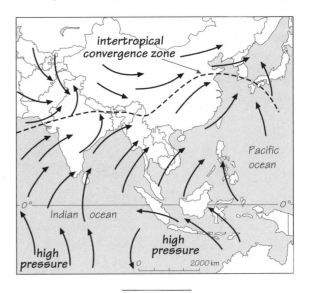

Summer monsoon in Asia.

kinetic energies of **rains** and, infrequently, with even more powerful impacts of hailstones—is the principal agent of continental **denudation** shaping the **Earth**'s surface.

Oceans

The Water would have been a more suitable name for the third planet than the **Earth: oceans** cover just over 70 percent of its surface to an average depth of 3.8 km. That this huge mass is such an outstanding regulator of the planet's energy balance is due to unique thermal properties of water.

This remarkable liquid has five major thermodynamic advantages: because of its capacity to form intermolecular hydrogen bonds it has an unusually high boiling point; its specific heat is 2.5–3.3 times higher than that of soils or rocks; its specific heat per volume (heat capacity) is roughly six times that of dry soils; its very high heat of vaporization makes it possible to transport a great deal of latent heat; and its relatively low viscosity makes it an outstanding carrier of heat in myriads of **ocean** eddies and in voluminous currents.

Not surprisingly, **oceans,** containing about 94 percent of all water, dominate the planet's energy balance. Some four-fifths of all **solar radiation** reaching the **Earth** enter the **atmosphere** above **oceans,** and low albedo of **oceans** (just around 6 percent) means that they absorb the incoming energy at a rate of about 65 PW, nearly twice as high as the total atmospheric absorption, and four times as high as the continental input. Inevitably, **oceans** also absorb most, about two-thirds, of the heat reradiated downward by the **atmosphere** and this raises their heating rate to about 175 PW.

In all but the shallowest **oceans** the air-sea interactions do not directly affect the whole water column. Cool and dark deep waters are isolated from the **atmosphere** by the mixed layer, a relatively shallow (a few meters to a few hundred meters) column that is usually made up of well

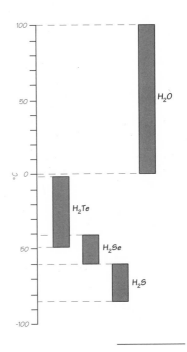

Boiling and freezing temperature of water compared to chemically similar compounds.

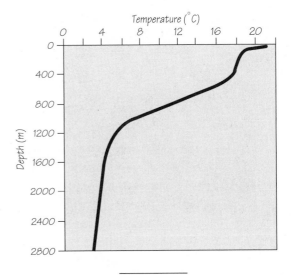

Typical thermal profile of ocean water.

churned-up *winds* and waves. In contrast to stable deep sea environment, the layer's temperatures have pronounced daily and seasonal fluctuations, although the high specific heat of water limits the range of possible amplitudes. Even relatively slight changes of *ocean* surface temperature can have profound climatic consequences: the recurring El Niño phenomenon, a tonguelike eastward expansion of warm surface waters whose effects reach as far as Canada and South Africa, is the best example of climatic teleconnections.

Because the conductivity of water is very low, the mixed layer transfers its energy downward largely by the mass motion of water in convective currents. These currents compensate for the extremely slow upwelling of deeper warm waters displaced by the equatorward movement of cold convective flows from the polar seas. In contrast to the gradual oceanwide upwelling, downward convection proceeds in restricted currents, including giant *ocean* cataracts. Perhaps the biggest one of these, flowing southward underneath the Denmark Strait between Iceland and Greenland, plunges about 3.5 kilometers and

carries 5 million m³/s, twenty-five times the flow of the Amazon.

Myriads of subsurface *ocean* eddies, often traveling for hundreds of km at various depths, also transfer considerable amount of energy—and salt. Perhaps the most remarkable combination of this transport is the formation of warm and salty eddies from the Mediterranean water flowing out of the Strait of Gibraltar. This warm but dense flow descends along the continental slope until it reaches neutral buoyancy at about one thousand meters. There it separates into clockwise-rotating lenses moving westward or southward for up to seven years before they decay or smash themselves on seamounts.

Global mapping of heat flux from the *ocean* surface to deeper water layers shows clear latitudinal maxima along the equator and along roughly 45°S in the southern Atlantic and Indian Oceans. This transfer is also substantial in some areas of coastal upwelling that experience vigorous convective exchange between surface waters and the deep *ocean,* most notably along the coasts of California and West Africa. But the upwelling regions associated

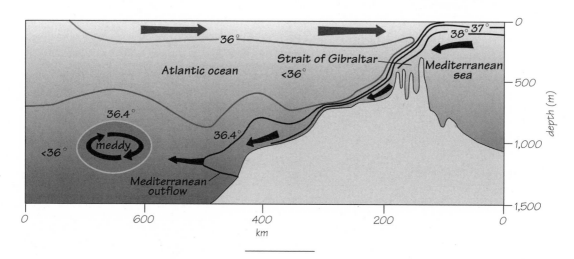

Eddies of warm, salty Mediterranean water flowing out of the
Strait of Gibraltar and drifting into the Atlantic.

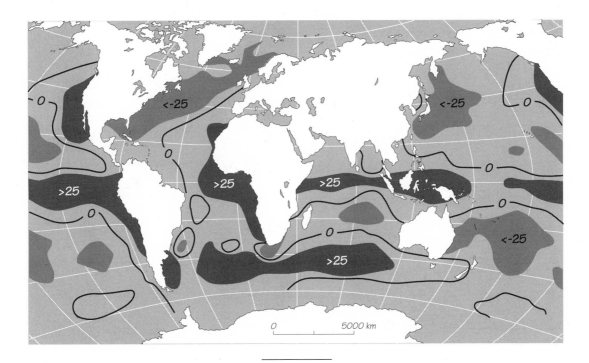

Heat flux from ocean surfaces to deeper water layers (W/m²).

with two major warm *ocean* currents, the Atlantic Gulf Stream and the Eastern Pacific Kuroshio, experience the opposite heat flux warming the *atmosphere.*

All upwelling regions—those along western shores of Americas, Africa, and India as well as the Western Pacific equatorial zone—are easily distinguished by high rates of *phytoplankton* productivity caused by substantial nutrient enrichment of what would be otherwise usual oligotrophic surface waters.

Radiation carries most (about four-fifths) of energy flowing from the mixed layer to the *atmosphere.* Latent heat moves nearly all of the rest in water vapor and *rains.* Quantifications of total latitudinal heat transport remain uncertain. The Atlantic Ocean transfers heat northward along its entire length, and the magnitude across the tropic is about 1 PW, a flux as high as that in the northern

Pacific. In the southern Pacific about 0.2 PW are moved poleward across the tropic, and its western part may be a major heat source for the southern Atlantic. So, probably, is the southern Indian Ocean for the Pacific.

Rivers

Total runoff of the *Earth*'s *rivers* averaged about 38,000 km³ during the 1980s, more than 5 percent below the mean of the late seventeenth century. Human interventions in water cycle, rather than a climatic change have been responsible for this decline. Potential energy of this water mass is roughly 300 EJ. Its release in kinetic energy of streamflows has been one of the most powerful geomorphic forces, although, compared to air speeds, even the fastest flowing waters move slowly. Typical stream velocities are only 0.5 meter per second in lowland *rivers,* and maxima in floods do not go higher than 2–3 meters per second. Vertical kinetic power densities of flooding streams will be thus less than 18 kW/m², the threshold for structural damage frequently surpassed by strong *winds.* But because of the longer durations of floods, and because of water's ability to weaken, or sweep away, the foundation soils, property destruction by flooding is often quite extensive.

Rivers do a great deal of geomorphic work not only through landscape erosion but also by transporting, sorting, and depositing the eroded materials. They carry three kinds of sediments produced by continental *denudation:* dissolved compounds from chemical weathering and atmospheric deposition; fine load of suspended clay and silt particles; and coarse bedload, ranging from gravel to boulders, requiring high velocities for downstream transport. Shares of the three components vary widely, but the ratio of 4:5:1 is an acceptable global mean. Because the stream competence—the maximum movable weight of individual pieces—varies with the sixth power of water velocity, a flow of four meters per second can carry stones sixty-four times more massive than the one of two meters per second. The total bedload stream capacity increases only with the cube of the velocity.

There are great differences among sediment loads carried by the world's major *rivers.* The Mississippi's sediment transport averages less than 70 t/km² of its drainage basin; the Amazon carries twice as much. The Huang He, the principal stream draining the world's largest loess plateau in northwestern China, has reached about 2000 t/km² during the 1980s. This was roughly a 25 percent increase in just three decades, a change caused by massive conversion of *grasslands* and *forests* to cropland.

Global estimates of sediment transported by *rivers* range from less than fifteen to more than thirty billion tonnes per year. Even the largest rate would represent a

A meandering river.

loss of less 0.1 percent of the overall potential energy change in the global streamflow. Deposition of sediments can lead to higher frequency of more damaging floods—but it has been also essential for creating alluvial plains whose rich soils supported the rise of first *traditional agricultures* in the Middle East and East Asia, and which are the most intensively cultivated areas on all continents.

Streams turning *waterwheels* also became the first practical sources of inanimate kinetic energy, and the nine-

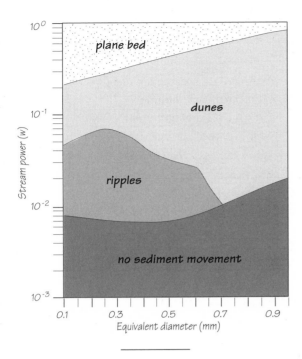

Relationship between stream power and bedforms.

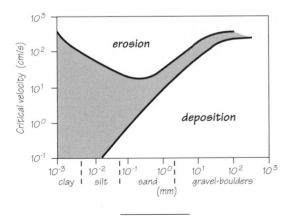

Stream-flow velocity and entrained particle size.

teenth century introduction of **water turbines** has made **rivers** convenient suppliers of inexpensive electricity. Theoretical global hydrogenerating capacity is almost exactly ten terawatts, but a large number of natural and engineering limitations (including flow fluctuations, competing water uses and impossibility to build dams in most lowland settings) reduces this potential by almost 90 percent.

Earth's Heat

Even if the **Earth** initially accreted cold, enormous amount of **heat** had to be lost during the subsequent formation of the liquid core, and periods of intensive volcanism; frequent impacts of massive objects further increased the surface heating. There is much less uncertainty about the **Earth**'s thermal history during the last three billion years: the planet has been cooling and a large part of its **heat** flow has been driving the global **geotectonics** creating new sea floor at oceanic ridges, a process accompanied by recurrent **earthquakes** and releases of lava, ash, and hot water by **volcanoes.**

There are only two possible sources of the **Earth**'s **heat** but a confident apportioning of their contributions remains elusive. Basal **heat,** released by a slow cooling of the **Earth**'s core, must account for a large share of the total flux, but calculations suggest that radioactive decay of ^{235}U, ^{238}U, ^{232}Th and ^{40}K could alone supply at least half, and perhaps up to nine-tenths of the planetary **heat** flux. Most of this disparity is caused by the uncertainty concerning the crustal concentration of ^{40}K. Whatever the actual breakdown, the total flux, based on thousands of measurements acquired since the 1950s, is around 40 TW.

Originally it was thought that continental and oceanic values are nearly equal, but they differ rather substantially. Regions of the youngest **ocean** floor may rate more than 250 mW/m^2, more than three times as much as the

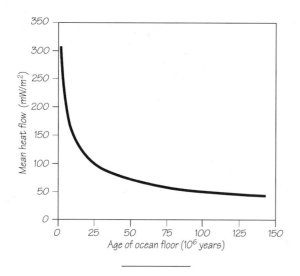

Heat flow as the function of the ocean floor's age.

youngest continental areas. Global mean for the sea floor is about 95 mW/m², more than 70 percent higher than that for the continental crust. Mean global flow of about 80 mW/m² is three orders of magnitude smaller than the average input of *solar radiation.*

The spatial distribution of *heat* flows reflects the age of crustal rocks and the intensity of geotectonic forces. The highest rates are in the eastern Pacific, coinciding with the areas of fastest sea floor spreading. Pinpoint maxima within these hot zones come from hydrothermal vents located along *ocean* ridges. Nicknamed black smokers because of the instant precipitation of black-pigmented sulfides in the stream, they can eject water with temperatures of up to 360°C and at rates of 25–330 megawatts; with fairly small vent openings these flows translate to power densities of 10^6–10^7 W/m². Such concentrated heat releases cannot be equaled by anything else but major volcanic eruptions.

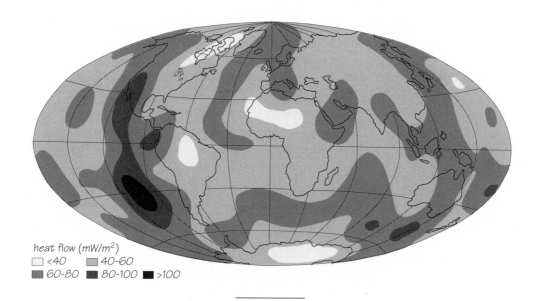

heat flow (mW/m²)

☐ <40 ▨ 40-60
▨ 60-80 ▨ 80-100 ■ >100

Large-scale averages of planetary heat flow showing the East Pacific maxima.

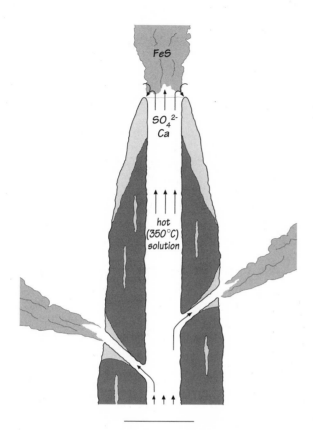

FeS

SO_4^{2-}
Ca

hot
(350°C)
solution

Black smoker, a deep ocean hydrothermal vent.

each for a fifth. By far the highest continental rates are in the southwestern part of North America. The lowest values, generally below 50 mW/m², are recorded over large regions of ancient Precambrian crystalline shields in Canada and Siberia.

Geotectonics

Grand features of the *Earth*'s surface—the planet's sea floor, its continents, and its mountain ranges—are generated by relentless movement of rigid blocks of lithosphere. Large oceanic plates diverge at the mid-ocean ridges where magma rises from the mantle to create new basaltic sea floor. Diverging plates slide about faults and eventually collide with continental margins where they are subducted into deep trenches to be recycled through the mantle. The journey between ridge and trench can be completed in just 10^7 years, while some parts of continents remain unusually stable, covered with rocks whose age is nearly twenty times the oldest sea floor's age of about two hundred million years.

Elsewhere collisions of relatively fast-moving oceanic plates with massive continents create steadily growing mountain chains. The most spectacular examples of subduction and mountain building are, respectively, the Pacific plate plunging into deep East Asian trenches, and the Himalayas rising from Indian plate's collision with Eurasia. In yet other segments of the lithosphere upwelling of hot mantle rock first weakens and then pierces the continental crust and eventually starts the rifting of continents by the creation of new *ocean* floor. The Red Sea, the Gulf of Aden, and the East African rifts demonstrate this process at its various stages.

Rifting appears to be fairly regular. Periods of intense compressive mountain building are discernible at intervals of four hundred to five hundred million years, followed about a hundred million years later by an upsurge in rifting, starting yet another round of supercontinental cycle

Cooler hydrothermal vents, releasing water at up to 30°C are the settings for unique ecosystems as based on primary production by chemoautotrophic *bacteria.* Similar to organisms common in sulfur-rich terrestrial environments, these microbes tolerate high acidities and derive energy from the oxidation of H_2S abundantly present in the vented water. Their chemosynthesis supports a surprising variety of invertebrate heterotrophs, including crabs, mussels, clams, and giant tubeworms. The last two species also harbor the chemolithotrophic *bacteria* as symbionts.

In total, the Pacific Ocean accounts for about half of the planetary heat flow, the Atlantic and Indian Oceans

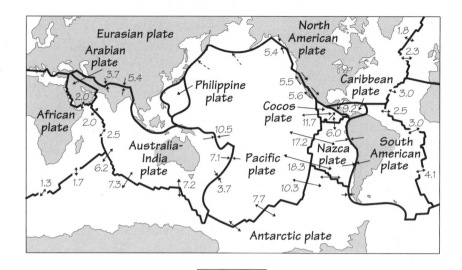

Tectonic plates, directions of their motion and average
annual speeds (in cm) of their travel.

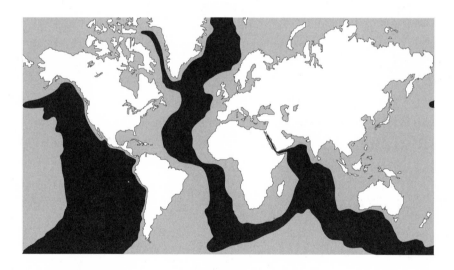

Area of the ocean floor created during the past 75 million years.

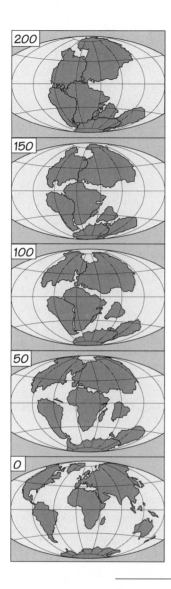

Breakup of a supercontinent formed about 300 million years ago and the subsequent continental drift (numbers are millions of years before the present).

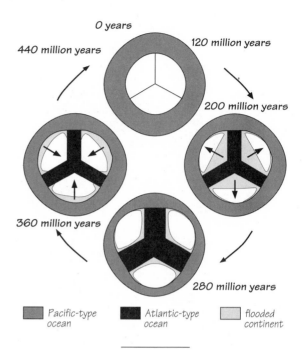

| Pacific-type ocean | Atlantic-type ocean | flooded continent |

Schematic illustration of the supercontinent cycle.

in which the breakup of land masses is followed by their eventual reassembly.

Magma plumes piercing the lithosphere also create many long-lasting hot spots associated with *volcanoes.* The Hawaiian islands and the chain of seamounts extending all the way to Kamchatka are the most spectacular manifestation of a vigorous mid-plate hot spot reasserting itself through a fast-moving Pacific plate, now through the continuing lava flows of Kilauea volcano and the slow creation of Loihi, Hawaii's future island.

Magma plumes so enormous that they had to originate in the lowermost layers of the mantle had repeatedly created large igneous provinces. The largest of these lava fields is the oceanic Ontong Java Plateau, covering about

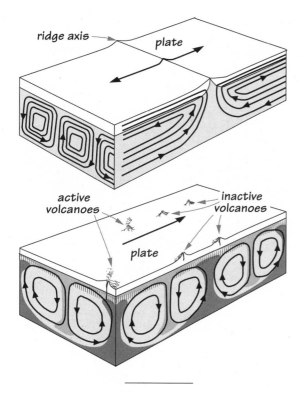

ridge axis — plate

active volcanoes — inactive volcanoes

plate

Models of mantle convection under tectonic plates.

meters, applies a torque on the viscous mantle and generates the main driving force for convection.

Its importance is demonstrated by the fact that plate velocities, which display no obvious relationship either with plate areas or with total lengths of ridges, correlate well with the total length of subduction zones. Plates with a large fraction of their boundaries in subduction zones have long-term velocities of sixty to ninety kilometers per million years, compared to speeds of less than forty kilometers per million years for plates without sinking slabs.

But the contribution of mantle upwelling is not negligible: it imparts considerable gravitational potential energy to relatively hot rocks over extensive areas: new seafloor created at *ocean* ridges is on the average at least three kilometers higher than the abyssal plains. Combination of "pull" along the subduction zones and "push" along the ridges can result in short-term speeds of nearly twenty centimeters a year for the fastest-moving plates. These include not only such fairly small slabs as Nazca and Cocos but also the huge Pacific plate, which means that mantle drag force, proportional to area and velocity, must be relatively small.

Measured flux of the *Earth*'s *heat* must be largely attributed to the formation of new oceanic lithosphere. With the mean global spreading rate of less than five centimeters a year and with *ocean* ridges totaling about 75,000 kilometers, some three square kilometers of new sea floor are formed, and subducted, every year. Change in *heat* content of formed and removed basalts implies the total annual *heat* loss rate of about 24 TW through the *ocean* floor. *Heat* loss through hot spots adds less than 3 TW for a grand total of about 27 GW — while the best estimates of overall oceanic *heat* flux are about 30 GW. Creation of new lithosphere thus accounts for about 90 percent of the *heat* loss through *oceans* and for about 60 percent of the *Earth*'s *heat* flux.

two million square kilometers. The Deccan Traps and Siberian Traps are the largest continental basalt formations. Creation of these vast igneous provinces affected the composition of the *atmosphere* by massive outgassing of CO_2 and SO_2: resulting higher tropospheric temperatures and acid *rains* had profound effects on biota.

Energetics of the *Earth*'s *geotectonics* is complex, and uncertainties exist even as far as the relative contributions of major forces are concerned. The two most important forces are associated with the convection of hot mantle material, and with the sinking of cold, negatively buoyant oceanic lithosphere along subduction zones. The latter process, produced by density differences which are greatest at depths of two hundred to three hundred kilo-

Earthquakes

Earthquakes originate overwhelmingly from grand-scale processes of the planetary *geotectonics* from creation, collision, and subduction of oceanic plates. No less than 95 percent of all tremors are concentrated along the edges of plates, and about nine-tenths of them are located in the Circum-Pacific Belt, where the relatively fast moving oceanic plates are either colliding with, or sliding past, the more massive continental plates. Most of the remaining *earthquakes* are associated with hot spots often marked by active *volcanoes.*

In aggregate, *earthquakes* account for only a very small fraction of energy liberated by the *Earth*'s tectonic processes. Since 1900 the seismic energy released annually by major tremors has averaged about 450 PJ, that is, no more than about 0.03 percent of the total flow of the *Earth*'s *heat.* Annual releases of seismic energy from all recorded *earthquakes* prorate to about 300 GW, and enlarging this aggregate by strain energy accumulated in irreversible deformations and by friction-generated heat along the faults could result in a grand total of about 1 TW, still no more than about 2.5 percent of the global *heat* flow.

But this global accounting tells little about the energy releases and the power of individual tremors. Most of them are too weak to be noticed by people, but every year there are major *earthquakes* causing a great deal of destruction: during the twentieth century they have been responsible for more death than floods, cyclones, and volcanic eruptions combined!

Energy of these tremors can be calculated either as the kinetic energy of the radiated seismic waves, or as the strain energy released by deformation of the ground. Either calculation has been rarely done directly. The most common approach has been to derive earthquake energies from the measures of magnitude, or of the moment, of tremors.

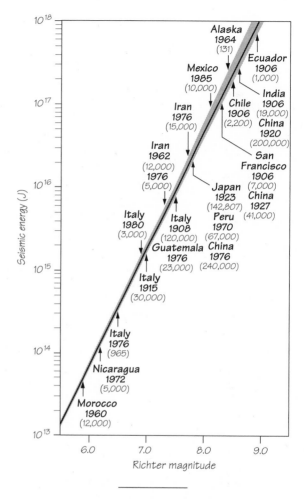

Plot of seismic energy and Richter magnitude of major twentieth-century earthquakes.

The standard magnitude measure, devised by Charles Richter in 1935, is the decadic logarithm of the largest trace amplitude (in micrometers) recorded with a standard torsion (Wood-Anderson) seismometer 100 km from the tremor's epicenter. In 1942 Richter published first correlations of magnitudes with total seismic energy releases, and his work has been followed by numerous modifications. Conversions follow the standard form $log_{10} E = a + bM$,

A building toppled during the Mexico City earthquake of 1985.

where E is the energy released as seismic waves (in ergs), M the Richter magnitude and a and b empirical coefficients ranging, respectively, from 6.1–13.5 and from 1.2–2.0. Alternative conversions are available for translating *earthquake* moment — a product of rigidity, average fault displacement, and average area displaced — into energy release.

The largest recorded *earthquakes* had Richter magnitudes of 8–8.9, and so their seismic energy releases amounted to between 48 PJ and 1.41 EJ. Uncertainties in estimating *earthquake* energies are well illustrated by comparing the published calculations for the 1906 San Francisco *earthquake*. Three strain method calculations give widely disparate values of 9, 40, and 175 PJ, and a single kinetic method value is only 2.5 PJ.

Because *earthquakes* are both very brief and spatially limited phenomena, they can produce extraordinarily high power ratings and densities. If a Richter magnitude 8 tremor lasted only half a minute, it would rate about 1.6 PW. If this power were evenly prorated over the area with radius of eighty kilometers, then its power density could be as high as 80 kW/m².

Obviously, such fluxes could be immensely destructive, but neither casualties nor material damage have any simple correlation with energy released by tremors. Residential and industrial densities and the quality of construction are far more important determinants of death toll and economic impact. For example, the death toll of the great 1923 Japanese *earthquake,* which set ablaze Tokyo's densely built wooden houses, was two hundred times higher than the casualties of San Francisco 1906 *earthquake,* which released roughly four times as much energy.

Forecasting these catastrophes remains elusive. There are some notable periodicities in the intensity of major tremor — between 1920 and 1990 energy released in great strike-slip and thrust *earthquakes* has had alternating cycles of twenty to thirty years — but our understanding is far from offering high-probability warnings.

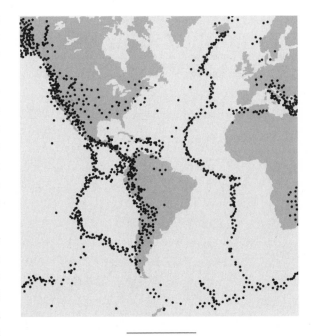

Earthquakes are located mostly along tectonic plate boundaries.

Underwater *earthquakes* generate seismic sea waves (tsunami) that can propagate for thousands of kilometers at speeds of 550–720 kilometers per hour with little loss of power. Virtually invisible at sea, tsunami rise to tens of meters only in shallow waters. The biggest ones can hit coastlines with as much as 200–500 MW/m² of vertical surface and generate horizontal impact power densities of 10^1–10^2 MW/m², far surpassing those of tropical cyclones and causing great damage and loss of life.

Volcanoes

The unmistakable association of active *volcanoes* with subduction zones of major tectonic plates locates most of recent eruptions around the Pacific Ocean, above all in Central and South America and in the Philippines, Japan, and Kamchatka. A less common category includes *volcanoes* associated with hot spots, places where tectonic plates are pierced by magmatic flows issuing from the mantle. The *volcanoes* of Hawaii and Central Africa are fine examples of this process. The best known historic eruptions include those of Greece's Thera (about 1500 B.C.), Italy's Vesuvius (79 B.C.), Japan's Fuji (A.D. 1707), Indonesia's Tambora (A.D. 1815) and Krakatau (A.D. 1883), and Washington State's Mount St. Helens in 1980. The last event became the most closely studied volcanic eruption to date. Consequently, we know not only the standard data about the volume of blast deposit (0.18 km³) and ejected fresh lava (0.5 km³), but also detailed breakdowns of energies involved in the eruption. Heat releases dominated the event: thermal energies of ejecta and avalanches, water and pyroclastic flows, and of the ash cloud added up to 1.66 EJ, about twenty times the total kinetic energy of the eruption.

Mount St. Helens' total energy release during nine hours of explosion on May 18, 1980, was about 1.7 EJ, implying average power of 52 TW, about five times the total of the world's annual primary commercial energy

Spectacular ash column created by the Mount St. Helens eruption.

consumption in the early 1990s. A number of well-monitored eruptions in the twentieth century were more powerful, most notably those of Kamchatka's Bezymyannyi in 1956 (3.9 EJ) and Japan's Sakurajima in 1914 (4.6 EJ). Energy released by the most powerful eruption of the nineteenth century, Tambora in 1815, was an order of magnitude higher: the best estimates based on the total volume of deposited ash point to a release over 80 EJ.

But even the largest historic eruptions are unremarkable compared to those which happened several hundred thousand years ago, and, in turn, the latter events seem minor compared with older magmatic releases. Among about ten relatively young calderas, enormous craters left behind the massive eruptions of the past one million years, are those of Yellowstone (formed about 600,000 years ago, seventy kilometers across, ejection of 1,000 km³ of pumice and ash) and Toba (in northern Sumatra, created 75,000 years ago, almost a hundred kilometers across, producing some 2,000 km³ of ejecta).

A prolonged period of volcanic eruptions that started 66 million years ago — several hundred thousand years of cataclysmic events which injected enormous quantities of dust into the *atmosphere* and produced more than

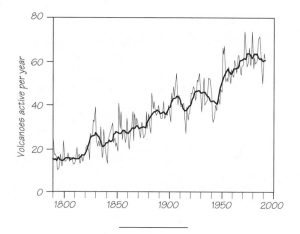

Total annual number of reported volcanic eruptions has doubled since 1800 (the thick line is the ten-year running mean).

2 million km³ of lava, creating India's massive Deccan Traps — seems to be at least as plausible a cause of mass extinction at the Cretaceous-Tertiary boundary as the *Earth*'s collision with an asteroid.

Historic eruptions have caused considerable loss of life (about 250,000 casualties since 1700) and a great deal of material damage, and they have been seen as major reasons of temporary global climatic change — but neither the loss of life and material damage nor any environmental consequences of eruptions are well correlated with their total energy release. Thermal releases are almost always dominant, surpassing all other energy flows by one to three orders of magnitude, but they can be concentrated in very different outflows.

In some eruptions the largest share of thermal energy is associated with ash clouds which rise into the stratosphere: ash from the Mount St. Helens eruption rose to twenty kilometers, ash from the 1991 eruption of Mount Pinatubo to over thirty kilometers, and Tambora's ash cloud reached up to forty-four kilometers. Spectacular sunsets and cooler weather in a year or two after such eruptions are well documented in some regions, but the

evidence for significant global cooling is mixed: 1816 was the year without summer in New England, but Tambora's ash had no cooling effect on Europe.

In other eruptions much of the thermal energy is carried by pyroclastic flows. These explosively produced currents combine volcanic ejecta ranging from μm to meters with hot gases. They can reach temperatures close to 1000°C, travel at speeds of up to three hundred meters per second and can extend as far as a hundred kilometers from their source vents. In 1902 these glowing clouds killed 28,000 people during Mount Pelée's eruption on Martinique. And in the case of Hawaiian *volcanoes* the outflow of *heat* is associated overwhelmingly with slowly moving lavas. Mauna Loa's 1950 eruption released nearly as much energy as did Mount St. Helens, but without any mudslides, avalanches, or ash clouds.

Spectacular or devastating as they are, volcanic eruptions account for a very small fraction of thermal energy

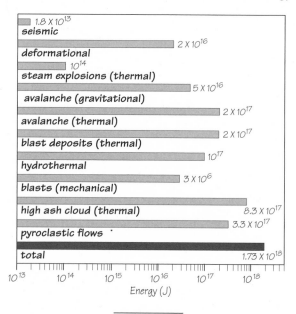

Kinds and magnitudes of energies released during the Mount St. Helens eruption between March 20 and May 18, 1980.

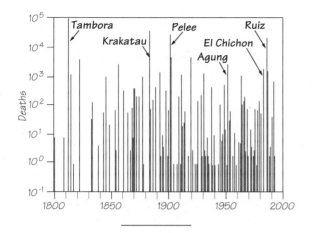

Death toll from volcanic eruptions, 1800–1990.

driving the **Earth**'s **geotectonics**. Assuming that continental lava flows are near 1 km³/year and that the seafloor eruptions add another 4 km³/year, the annual global **heat** loss due to **volcanoes** would be about 800 GW, only about 2 percent of the Earth's total geothermal flux.

Denudation

Powerful geomorphic forces can act swiftly. Gravity-induced mass wasting can move huge volumes of soil or stones in devastating landslides and rockfalls in a matter of seconds. During heavy **rains** gully and channel erosion can rearrange fields, riverbanks, and floodplains in hours or days, and hurricane-driven **winds** can have a similarly rapid effect on coastlines. And surfaces can be remodeled almost instantaneously by **earthquakes** and **volcanoes**. But **denudation** of most of the continents — the process of weathering and the subsequent stripping of weathered materials by erosion — is a gradual change whose rates are frequently so slow that a lifetime of observations yields no obvious alterations.

The Bubnoff unit (B) — **denudation** of one millimeter in a thousand years (or 1 μm/year) — is a convenient measure of this change. Dissolution by precipitation can reduce hard igneous or metamorphic rocks by just 0.5–5 B, limestones by up to 100 B. **Denudation** in dry lowlands usually proceeds at rates no higher than 10–15 B, in humid tropics at 20–30 B. Changes in mountainous terrain can be much faster, with rates up to 800 B in regions with relatively rapid moving massive glaciers (Southeast Alaska) and up to nearly 10 kB in the youngest zones of continuous uplift (the Nanga Parbat section of the Himalayas). But even relatively rapid **denudation** rates result from the application of modest forces. An environmentally, and economically, important example illustrates this uniformly low power of geomorphic processes. Without erosion agricultural soils would be deeper, but their leached top layers would become nutrient-poor. Weathering and erosion replenish minerals in the root zone and help maintain soil fertility — but not if the process moves too fast.

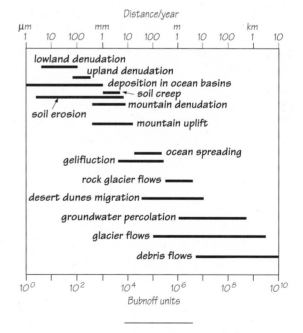

Intensity of denudation for major geomorphic processes; mountain uplift and ocean spreading are included for comparison.

A glacier in the Sikkimese Himalayas.

Maximum annual topsoil loss compatible with sustainable cultivation of crops is about eleven tonnes per hectare for most of America's farmland. About two-fifths of the country's fields are now eroding at faster rates, and the national average for water erosion alone is almost ten tonnes per hectare, or (assuming topsoil density of 1.8 t/m³), about 550 B. If, after a mean drop of about 250 m,

all of the water-eroded topsoil ended up as silt in neighboring *oceans,* the mass removed from 170 million hectares of American farmland would have its potential energy reduced by about 4 PJ. This implies minuscule geomorphic power density of less than 80 μW/m².

The dominant role of *rains* in the *denudation* process becomes obvious when comparing the kinetic velocity of raindrops with that of surface runoff. The largest raindrops, with diameters of five to six millimeters, have terminal velocities of nine meters per second, and kinetic energy of their impact will be roughly forty times their mass. Even if as much as one-half of this precipitation became surface runoff with a mean speed of one meter per second, its kinetic energy of flowing water would be equal just to a quarter of the transported mass.

Consequently, raindrop erosion would be two orders of magnitude (in this particular case 160 times) more

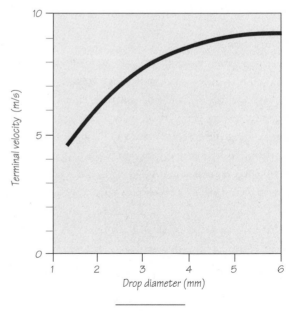

Terminal velocity of raindrops.

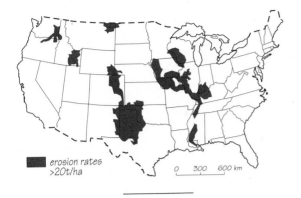

erosion rates
>20t/ha

0   300   600 km

Areas of the highest soil erosion in the United States.

powerful than the surface runoff. But other variables are also of great importance, none more than the extent and the kind of vegetation cover. Cultivation of row crops, leaving much land exposed to direct raindrop impact before the closure of plant canopies, is particularly conducive to high erosion rates: in highly mobile soils they may be more than 10,000 times higher than in *forest* soils, protected not only by overlapping needles and leaves but also by a thick litter layer.

Planetary total of denudational energy can be calculated by assuming removal of at least 50 B of material with average density of 2.5 g/cm³ (125 t/km²) and the mean continental elevation of 850 m. Potential energy of the *Earth*'s landscapes would be then reduced annually by about 135 PJ. This is a tiny flow compared to other planetary energy fluxes: at 4.3 GW its total is less than 0.05 percent of potential energy lost in precipitation runoff, only about 0.01 percent of the *Earth*'s *heat,* and it is equal to less than 2 x 10⁻⁷ of *solar radiation* absorbed by continental surfaces. Clearly, only a negligible fraction of *solar radiation* — be it directly as sunlight or indirectly as running water and *wind* — energizes the denudation of continents.

Moreover, the counteracting forces have been negating this slow change. Without continuing tectonic uplift even alpine ranges four thousand meters high would be leveled in less than five million years when subjected to *denudation* rates of 1–5 kB — but they are actually an order of magnitude older. Uplift rates of 5–10 kB are fairly common, and many regions are rising at more than 20 kB, up to ten times faster than their rate of *denudation.* But in mountain belts that intersect the snowline, receive abundant precipitation, and are extensively glaciated, it appears that their overall height is limited by the rapid rate of *denudation* rather than by the process of tectonic uplift. The northwestern Himalayas, including the famous Nanga Parbat area, are a perfect example: although many of its peaks surpass seven thousand meters, only one percent of the land is above six thousand meters.

Space Encounters

Even the most jarring temporary intensifications of the *Earth*'s normally much more sedate geotectonic and atmospheric energy flows — be they caused by extraordinarily powerful *volcanoes* and *earthquakes* or by unusually damaging *winds* and *rains* — appear trifling in comparison with the planet's repeated collisions with relatively massive extraterrestrial bodies. The *Earth* is constantly showered with myriads of unseen microscopic dust particles abundant throughout the solar system, and even bits with diameter of 1 mm, large enough to leave behind a light path as they self-destruct in the *atmosphere,* come every thirty seconds.

Collisions with meteorites are also relatively frequent as solids with 1 m diameter strike the planet at least once a year, but even the impacts of larger chunks have only localized effects. This is because all meteorites wandering away from the asteroid region between Mars and Jupiter are on circumsolar paths in the same direction as the

Earth, and their impact velocities are less than fifteen kilometers per second. The almost perfectly symmetrical Arizona crater was created 25,000 years ago by a meteorite impacting at eleven kilometers per second. Its impact power was about 700 PW.

But even such huge energy releases pale in comparison with a head-on encounter with a typical comet. Its mass (at least five billion tonnes) and its relative speed (up to 70 km/s) would produce kinetic energy of about 1 x 10^{22} J. Even if all but 10 percent of it were dissipated in the *atmosphere* the impact energy of 1 x 10^{21} J would

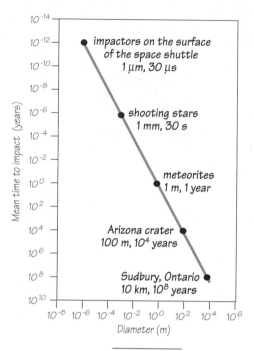

Frequency of terrestrial impacts declines exponentially with the size of objects.

"And the stars of heaven fell onto the earth . . ." :
Albrecht Dürer's illustration of a line in Revelation.

equal the explosion of about 2500 100-Mt hydrogen bombs. That such an event would have awesome climatic consequences is clear; that a similar impact caused the Cretaceous extinctions is disputable. Given their relatively long span, a period of intensive volcanism is a more plausible explanation.

Collisions with comets should occur roughly every fifty to sixty million years, which means that the biosphere, evolving for nearly four billion years, had to cope with some two scores of such events. Clearly, it was successful in doing so inasmuch as these collisions may have modified it but could not terminate it. Nor has been the biosphere's evolution derailed by high-energy bursts from nearby supernovas. With the galactic mean of a supernova explosion every fifty years, the solar system is within a

Arizona's famous Barringer Meteoric Crater is
1220 meters in diameter and about 180 meters deep.

Orbits of some major comets.

hundred parsecs of such an event every two million years, and within ten parsecs only once every two billion years. In the latter case an exploding supernova would flood the top of the *atmosphere* with an x-ray and very short UV flux about ten thousand times higher than does the incoming *solar radiation,* and the *Earth* would receive an equivalent of its annual dose of ionizing *radiation* in just a few hours.

Exposures to five hundred roentgens are lethal to most vertebrates, but in spite of up to ten postulated episodes of this magnitude during the past five hundred million years there is no obvious record of such effects on biospheric evolution. And if it is assumed that an advanced civilization could prepare shelters for its population during the year elapsing between the arrival of light and cosmic *radiation,* then an unavoidable dose of five hundred roentgen should be encountered only about once in a billion years, a period permitting the development of a society whose knowledge might defend it against even such a rare flux.

2
Plants and Animals

Both autotrophs (photosynthesizers) and heterotrophs (decomposers and *animals*) use *adenosine triphosphate* as their principal energy currency. Single-celled *Archaea*, formerly classed among *bacteria*, reduce simple inorganic compounds. *Bacteria* have many species in both trophic groups, and play indispensable roles in biospheric cycles maintaining essential nutrient flows needed in *photosynthesis*. C_3 and C_4 *plants* are dominant autotrophs (cacti fix carbon by following yet another photosynthetic pathway), and *forests* and *grasslands* are the biosphere's most important terrestrial biomes whose standing phytomass greatly surpasses that of oceanic *phytoplankton.*

Energy supply for *heterotrophic metabolism* depends, directly or indirectly, on eating virtually any available phytomass tissues. Specialized *herbivores* can even digest abundant but resilient cellulose and lignin. *Carnivores* have smaller amount of biomass to prey on, but their diet is highly nutritious and easily digestible. Differences in climate, plant productivity, and heterotrophic biodiversity produce *food chains* ranging from single links to complex webs.

Endotherms (birds and mammals regulating their body temperature) have unmatched environmental adaptability, but *ectotherms* (organisms whose temperature changes with that of the environment) are much more abundant, both as species and as individuals. Differences between these two modes of heterotrophic existence extend to energy costs of *reproduction* and *growth*—and to their locomotion. Although there are many very good ectothermic swimmers and some good runners and fliers, the best performances in *swimming,* and to an even greater extent in *running and jumping* and in *flying,* belong to *endotherms.*

Adenosine Triphosphate

Adenosine triphosphate (ATP) provides the key link between cellular catabolism (degradation of nutrient substrates) on one side, and anabolism (biosynthesis of complex compounds), locomotion (muscle contraction), and active transport of metabolites in *bacteria, plants,* and *animals* on the other.

ATP is formed from adenine, ribose, and a triphosphate tail.

Formed during the respiration of glucose, this organic phosphate became the principal cellular energy currency because of its intermediate free energy of hydrolysis (−31 kJ/mole): at the beginning of glycolysis it can readily donate a phosphate group to produce glucose-6-phosphate (with free energy of just −13.8 kJ/mole) — or it can be easily formed from a much more exergonic (−49.3 kJ/mole) 1,3-diphosphoglyceric acid later in the process.

During the first step of these enzymatically catalyzed reactions glucose or glycogen are broken down to pyruvic acid. In an anaerobic environment this pathway ends in either lactic acid (causing tired muscles) or in *ethanol* and CO_2 (in bacterial fermentation). Where oxygen is available it proceeds to tricarboxylic acid cycle, converting various organic compounds to CO_2, transferring the released electrons, reducing oxygen to water and producing large amounts of *ATP*.

A large amount of energy is released when ATP is hydrolyzed by water to adenosine diphosphate and phosphate ion.

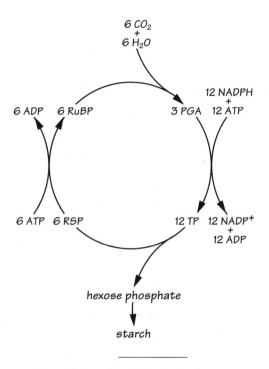

6 CO$_2$
+
6 H$_2$O

6 ADP   6 RuBP   3 PGA

12 NADPH
+
12 ATP

6 ATP   6 RSP   12 TP

12 NADP$^+$
+
12 ADP

hexose phosphate

starch

ATP and ADP in photosynthetic cycle, the most important
reduction reaction in the biosphere.

In prokaryotic cells the maximum energy gain from this complex sequence is thirty-eight moles of **ATP** for each mole of glucose, an overall free energy gain of about 2.8 MJ. With 31 kJ/mole released from each **ATP** transformation to adenosine diphosphate, the overall efficiency of the whole sequence would be about 40 percent. In eukaryotic cells, where up to 50 kJ/mole can be released during the transformation of **ATP**, the overall efficiency may be over 60 percent.

Respiration of fatty acids has just about the same efficiency, but the compounds have much higher energy density than glucose. Maximum efficiencies for the breakdown of glucose to lactic acid are only about 31 percent, and only invertebrates living in oxygen-poor environments rely on this process for extended periods of time.

Stunning is the only appropriate adjective describing the intensity of **ATP** generation. Humans synthesize and break down about three grams of **ATP** for every gram of dry body weight, producing and using daily more of the phosphate than is their total body mass! But this turnover is quite unimpressive compared to respiring *bacteria*. *Azotobacter,* a common soil species forming ammonia by using atmospheric **nitrogen,** produces seven kilograms of **ATP** for every gram of its dry mass!

High intensities of *heterotrophic metabolism* mean that living organisms surpass the *Sun* in power output per unit of mass. Given the star's enormous mass (1.99 x 10^{33} g), its immense solar luminosity (3.9 x 10^{26} W) prorates to just about 200 nW/g of the stellar matter. In contrast, the daily *metabolism* of children (averaging around 3 mW/g of body weight) proceeds at a rate about 15,000 times higher, and respiring *bacteria* reach up to 100 W/g, or 500 million times the *Sun's* rate. Stars astonish with their total energy fluxes, but **ATP**-driven energy conversions in heterotrophic organisms have unrivaled intensities of energy conversions per unit mass.

Archaea and Bacteria

Recent genetic studies relying on amplifying and cloning ribosomal RNA sequences, an approach pioneered by Carl Woese, have given us a new classification of living organisms — and a new domain of life, **Archaea.** Under the microscope these single-celled microorganisms are indistinguishable from *bacteria,* but genetically they are quite distinct and are actually more closely allied to *Eucarya,* life's third domain, which includes plants and animals.

Today's Archaea are the descendants of the oldest living inhabitants of the biosphere. Origins of methanogens, reducing CO$_2$ or other very simple sources of carbon (formate, methanol, or acetate), and sulfate-reducing archaeons, go back nearly four billion years. Many of these anaerobic organisms continue to thrive in extreme

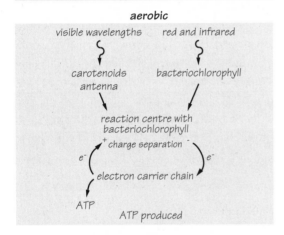

anaerobic

550-660nm

phycobiliproteins antenna

H_2O

reaction centre II chlorophyll

charge separation

660nm

chlorophylla

NADPH

reaction centre II chlorophyll

$1/2 O_2$

e^-

electron carrier chain

e^-

NADP

e^-

ATP and reducing power produced

aerobic

visible wavelengths

red and infrared

carotenoids antenna

bacteriochlorophyll

reaction centre with bacteriochlorophyll

charge separation

e^-

electron carrier chain

e^-

ATP

ATP produced

Photosynthetic pathways of anaerobic and aerobic bacteria.

by far the most abundant organisms in the biosphere, and they possess a great metabolic plasticity, including both anaerobic and aerobic species and autotrophs as well as fermenting and respiring heterotrophs. The earliest bacterial primary producers were chemoautotrophic, synthesizing new biomass from inorganic compounds without **solar radiation** by oxidizing mostly **nitrogen** or sulfur compounds.

Again, these organisms remain indispensable, most notably for running the two key biogeochemical cycles. In the **nitrogen** cycle they derive their energy by converting NH_3 to NO_2 (*Nitrosomonas*) and NO_2 to NO_3 (*Nitrobacter*). In the sulfur cycle they energize themselves by oxidizing sulfides, thiosulfates, and sulfites and producing sulfates. In turn, they are specialized anaerobic sulfate-reducing **bacteria** producing H_2S or elementary sulfur.

Bacteria deriving their energy from **photosynthesis** include organisms producing new biomass anaerobically and in two modes under aerobic conditions: the first one, discovered during the 1970s, does so anoxygenically; the other, more common one involves oxygenic **photosynthesis.** The anaerobic phylum includes purple sulfur and terrestrial and oceanic environments—*Methanococcus* survives on the sea floor under pressure of up to more than two hundred atmospheres; *Pyrodictium,* associated with deep-sea hydrothermal vents, reproduces in waters at 80–110°C; *Archaeoglobus* thrives in deep oil wells—but other archaeons are common in farms soils.

All higher forms of life depend on **Archaea** and **bacteria:** They generate most of the flows in the grand biogeochemical cycles of carbon, **nitrogen,** and sulfur, circulating the elements of life. Not surprisingly, they are

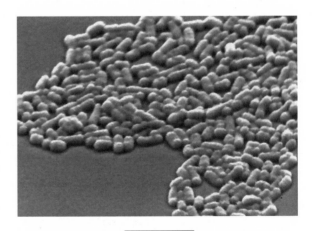

A cluster of rod-shaped *Escherichia coli* bacteria, a common human pathogen.

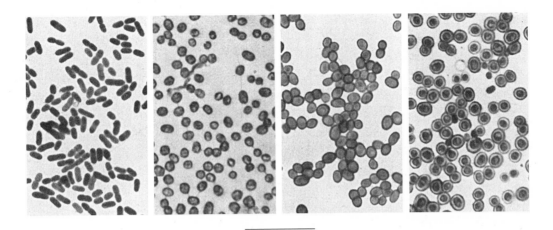

Growth stages of *Azotobacter vinelandii* from
Winogradsky's classic *Microbiologie du sol*.

green sulfur ***bacteria*** common in anoxic muds and using
either H_2S or H_2 as electron donors to reduce CO_2. The
second sequence is carried on above all by ***bacteria*** in various marine environments. The third catergory is made up
of cyanobacteria.

These organisms, formerly called blue-green algae,
synthesize new phytomass much like real algae or higher
plants. Abundant in virtually all aquatic environments,
they are important constituents of ***phytoplankton*** and are
typically responsible for 5–20 percent of all oceanic ***photosynthesis***. Plasticity of cyanobacteria is demonstrated in
three remarkable ways. In the presence of nutritious substrates, many species can become heterotrophic; in aquatic
ecosystems they can shift between oxygenic and anoxy-
genic ***photosynthesis*** (relying on sulphide oxidation),
clearly demonstrating the evolutionary continuum of trophic strategies; other species are symbiotic with algae,
fungi, higher plants, and animals.

Notable respiring heterotrophs include omnibacteria
associated with plant, animal, and human diseases and ac-
tinobacteria and pseudomonads, including decomposers
that are either specialized for feeding on particular com-

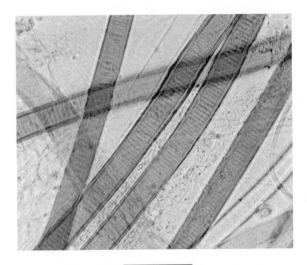

Filaments of *Lynbya aestuarii*, a cyanobacterium
forming dense mats in benthic waters.

pounds (such as *Cellulomonas* thriving on polysaccharides)
or can use just about any carbon substrate (such as the
ubiquitous motile *Pseudomonas*). Heterotrophic ***bacteria***
also make up a substantial share (10–40 percent) of plank-
ton in both coastal and open ocean waters. Anaerobic

fermenting heterotrophs are implicated in infectious diseases ranging from dental abscesses to edocarditis. Soil *bacteria* provide a number of irreplaceable ecosystemic services mostly connected with the carbon, *nitrogen,* and sulfur cycles. Poor soils may contain fewer than a hundred kilograms of *bacteria* per hectare — but up to seven tonnes per hectare can be found in the rhizosphere of alfalfa and another two tonnes per hectare outside of it.

Large shares (commonly over 70 percent) of soil *bacteria* are dormant in order to survive in nutrient-poor environments. Bacterial *phytoplankton* in nutrient-poor waters often resort to rapid fragmentation into ultramicrocells in response to starvation. Many *bacteria* also do not behave as solitary cells but rather much like components of multicellular organisms. They differentiate into various cell types and form complex communities that move along marked chemical trails and engage in group predation.

Photosynthesis

The well-known basic equation describing the endothermic reaction requiring 2.8 MJ of *solar radiation* to synthesize one molecule of glucose from six molecules of CO_2 and H_2O is an oversimplified black box. A more realistic black box looks like this: $106\ CO_2 + 90\ H_2O + 16\ NO_3 + PO_4$ + mineral nutrients + 5.4 MJ of radiant energy = 3,258 g of new protoplasm (106 C, 180 H, 46 O, 16 N, 1 P, and 815 g of mineral ash) + $154\ O_2$ + 5.35 MJ of dispersed heat. Without macro- and micronutrients there would be no new phytomass whose bulk is made up of basic *nutrients* needed by all heterotrophs: complex sugars, fatty acids, and proteins.

A look into the box should start with light-sensitive pigments dominated by chlorophylls. They absorb *solar radiation* in two narrow bands, one between 420 and 450 nm, the other one between 630 and 690 nm: *photosynthesis* is energized by blue and red light, and photosyn-

thetically active radiation (PAR) accounts for slightly less than one-half of total insolation. This energy is not needed for CO_2 reduction, but rather for the regeneration of compounds consumed during the fixation of the gas.

Synthesis of phytomass by the reductive pentose phosphate (RPP) cycle — a multistep pathway of enzymatically catalyzed carboxylation, reduction, and regeneration — must be preceded by the formation of *adenosine triphosphate* (ATP) and nicotinamide adenine dinucleotide phosphate (NADPH), the two compounds energizing all biosynthetic reactions. Synthesis of three molecules of ATP and two molecules of NADPH needed to reduce each molecule of CO_2 will require ten quanta of *solar radiation* with wavelengths around the red absorption peak of chlorophyll (680 nm). Carbon from CO_2 will combine with hydrogen from water and with micronutrients to produce new phytomass containing about 465 kJ/mol.

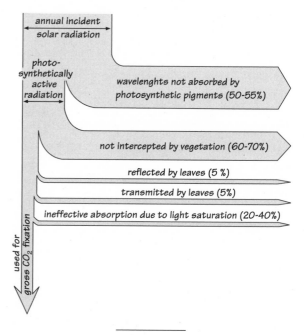

Average partitioning of solar radiation reaching photosynthesizing plants.

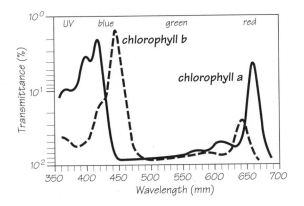

Absorption of light by chlorophylls.

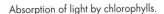

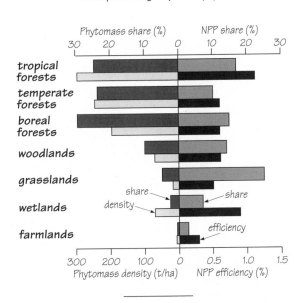

Phytomass shares and densities and NPP shares
and efficiencies for major ecosystems.

energy of nearly 176 kJ. Assuming that PAR is roughly 45 percent of the direct light the overall efficiency of *photosynthesis* would be about 11 percent (465/[1,760/0.43]). Reflection of light by leaves and its transmission through the canopies reduce this rate by at least one-tenth.

No *plant* comes even close to this theoretical peak. Part of the light absorbed by chlorophylls (commonly 20–25 percent) will be reradiated as heat because the pigment cannot store the sunlight and enzymatic reactions cannot keep up with the incoming energy flux.

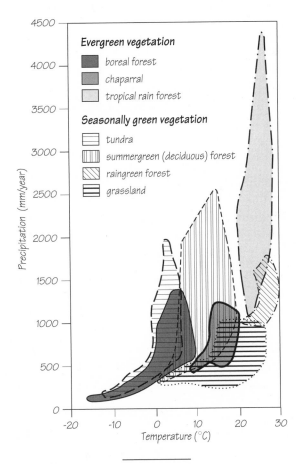

Major biomes delimited by precipitation and temperature.

Energy content of a quantum of red light would be 2.92×10^{-19} J (Planck's constant, 6.62×10^{-34} multiplied by the light frequency, a quotient of the light speed and the mean wavelength). One einstein (that is one mole, or Avogadro's number, 6.02×10^{23}) of red photons has an

Respiration dissipates fixed carbon in *plant* metabolism and in the maintenance of supporting structures. Its rates for individual species are determined primarily by their photosynthetic pathway: The differences between C_3 *and* C_4 *plants* are substantial. Respiration on community and ecosystemic scale depends on the growth stage: it ranges from less than 20 percent in young, rapidly growing *plants* to over 90 percent in mature *forests.*

With 25 percent each for reaction and respiration losses, photosynthetic efficiency would be a bit above 5 percent. Only here do theoretical calculations and actual performances meet: the highest recorded field values of net *photosynthesis* in some highly productive *plants* under optimum conditions during short periods of time are 4–5 percent. For most *plants* the rates are further limited by shortages of nutrients, particularly *nitrogen,* and water, for those in higher altitudes and latitudes by low temperatures. Best rates for highly productive natural formations (especially wetlands) and crop fields are 2–3 percent.

On ecosystemic scales conversions range from about 1.5 percent for tropical and temperate marshes and temperate *forests* to just around 0.1 percent for very arid *grasslands.* Annual production of at least 100 billion tonnes (about 2000 EJ) of new phytomass implies a global terrestrial mean of about 0.6 percent. *Photosynthesis* in oceans, severely stressed by nutrient shortages, is even less efficient: average productivity of just over 3 MJ/m² implies photosynthetic efficiency of mere 0.06 percent. The weighted global mean would be then just around 0.2 percent: merely one out of every 500 quanta of solar energy reaching the *Earth*'s surface is actually transformed into biomass energy in new *plant* tissues.

Most of this energy is stored in simple sugars and their polymers (especially in cellulose) and in lignin; more energy-dense *lipids* are usually present only in seeds. Using the best areal estimates and typical ecosystem storages results in a global estimate of just over one trillion tonnes

of standing phytomass for the early 1990s. This prorates to roughly 160 MJ/m² of land. Most of this phytomass is stored in *forests,* a much smaller fraction is in *grasslands,* and less than half a percent of the global total is in agricultural crops.

C_3 and C_4 Plants

Less than a decade after Andrew Benson and Melvin Calvin unraveled the basic sequence of RPP cycle of *photosynthesis,* Hal Hatch and Roger Slack discovered that many *plants* follow initially a different fixation path. Unlike in the RPP cycle—where the reduction of CO_2 produces phosphoglyceric acid, a compound containing three carbon atoms—these *plants* first synthesize oxaloacetate in their mesophyll cells. This four-carbon acid is first reduced to malate, another four-carbon compound, and then it is moved into chloroplasts of the bundle sheath cells where CO_2 is regenerated and the released pyruvate is returned to the mesophyll. The regenerated CO_2 then enters the Calvin cycle and its carbon is incorporated into new phytomass.

Because of the initial reduction of CO_2 into malate, C_4 *plants* have very low internal concentrations of the

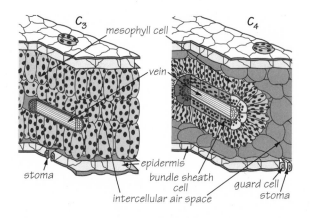

Leaf sections showing anatomical differences between C_3 and C_4 plants.

characteristics	C_3 plants	C_4 plants
light saturation	saturation at <300 W/m^2	no saturation
maximum rate of net photosynthesis (mg CO_2/dm^2 leaf area/h)	15 - 35	40 - 80
optimum temperature for net photosynthesis (°C)	15 - 25	30 - 45
CO_2 compensation concentration in photosynthesis (ppm CO_2)	30 - 70	0 - 10
photorespiration	high	low
transpiration rate (g H_2O/g dry wt)	450 - 950	250 - 350
growth rate (g dry wt/dm^2 leaf area/d)	0.5 - 2	3 - 5

Principal differences between C_3 and C_4 plants.

gas—close to just 10 ppm, compared to 350 ppm in the ambient air—and they are much more efficient users of water. Given the facts that water-vapor pressure is so much higher inside the leaves than in the atmosphere, and that these pressures are two orders of magnitude higher than the differences between concentrations of ambient and leaf CO_2, *photosynthesis* will always entail huge H_2O exports in exchange for relatively small CO_2 imports. But the most efficient C_4 species can keep this lopsided trade down to 450–600 moles of H_2O for every mole of CO_2, while common rates for C_3 *plants* are around 1000, and they may be as high as 4000.

Another critical advantage of C_4 pathway is virtual absence of photorespiration and oxidation of the synthesized phytomass. In C_3 *plants* all chloroplasts are in mesophyll cells, whereas in C_4 species they are in a bundle sheath layer of cells surrounding vascular conducting tissues. Consequently, in C_4 *plants* ribulose 1,5 bisphosphate carboxylase—an enzyme that can use O_2 as well as

CO_2 and hence also produce rather than only reduce the gas—is present only in the bundle sheath cells where O_2 levels are naturally low, and possibilities for oxidation are negligible.

Regeneration of pyruvate needs additional energy from **adenosine triphosphate,** but with negligible photorespiration, and with pyruvate carboxylase having a greater affinity for CO_2 than ribulose carboxylase, net photosynthetic efficiencies of Hatch-Slack pathways are substantially higher—up to 70 percent when averaged over the whole growing season—than in C_3 *plants.*

Differences in water needs and primary productivity have enormous economic implications for cultivation of crops—and they explain why corn has become the leading global feed *grain* species, or why production of sugar from beets cannot be more efficient than its extraction from cane. Many C_4 species—such as big bluestem, bermudagrass, paragrass, and buffelgrass—also contribute greatly to productivity of *grasslands.* Unfortunately, the two most preferred food *grains,* wheat and rice, are both C_3 *plants,* as are all leguminous species high in protein.

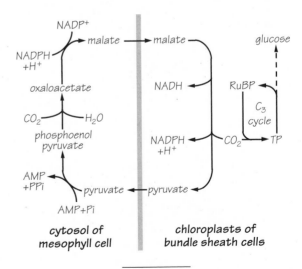

Basic sequence of the C_4 photosynthetic cycle.

C_4 *plants* have yet another, climatic, advantage. While *photosynthesis* in C_3 *plants* saturates with *solar radiation* around 300 W/m² and proceeds optimally at 15–25°C, C_4 species have no light saturation and do best in temperatures of 30–45°C. C_4 *plants* thus have a competitive advantages in sunny, warm climates.

Phytoplankton

Most of the terrestrial phytomass productivity and storage is concentrated in large trees in *forests*—but *phytoplankton* species, the principal producers in the *ocean,* are tiny unicellular drifters. They come in sizes between less than 2 and 200 μm in diameter, and include varying proportions of *bacteria* and eukaryotic protoctists. The coccoid cyanobacteria are so common in some oligotrophic waters that they may account for the largest share of phytoplanktonic production.

Photosynthesizing protoctists range from the smallest pigmented flagellates (such as cryptomonads and

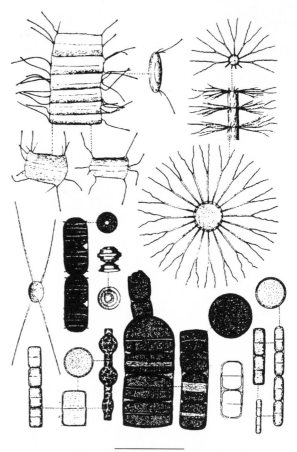

Wispy *Chaetoceros* and segmented *Melosira* diatoms.

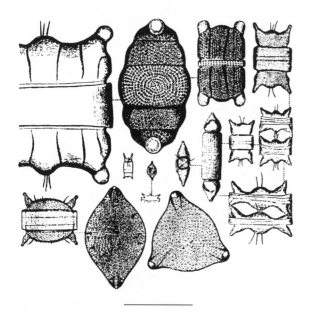

Bulky diatoms of the genus *Biddulphia.*

chrysophytes) to larger (more than 10 mm), and usually dominant, diatoms and dinoflagellates. Diatoms have nonflagellated cells with walls of amorphous silica mixed with organic compounds. Both the centric (radially symmetrical, dominant in oceans) and pennate (bilaterally symmetrical) diatoms have a wide variety of intriguing designs and they often form long chains. So do the dinoflagellates, a phylum that also contains many nonphotosynthesizing species.

Phytoplanktonic productivity is controlled by water temperature and by the availability of *solar radiation* and nutrients. Temperature is rarely the decisive factor: many species are highly adaptable and have very similar productivities in very different environments. Adaptations to different solar inputs are also common, but both the volume of diatom cells and their chlorophyll content increases with higher light intensity. Nutrient supply presents the greatest challenge in most open waters: *nitrogen* is the most important limiting macronutrient, but phosphorus shortages are also common, as are those of some key micronutrients, above all iron and silica.

Surfaces of open *oceans,* and the waters layers immediately underneath, are among the least productive environments on the *Earth.* The richest concentrations of macronutrients in open waters are between five hundred and one thousand meters, well beyond the euphotic zone, the topmost water layer penetrated by light, which goes to about one hundred meters in the clearest seas. Tiny size of dominant primary producers is an effective adaptation to this nutrient scarcity: high surface/volume ratios and gentle sinking of *phytoplankton* cells through the euphotic layer increase the rates of nutrient absorption.

Only where upwelling enriches the surface layers by imports of cool, nutrient-laden water can *phytoplankton* productivity soar. Waters off Peru, California, northwest and southwest Africa, and western India are prominent examples of coastal upwelling; the mid-Pacific near the

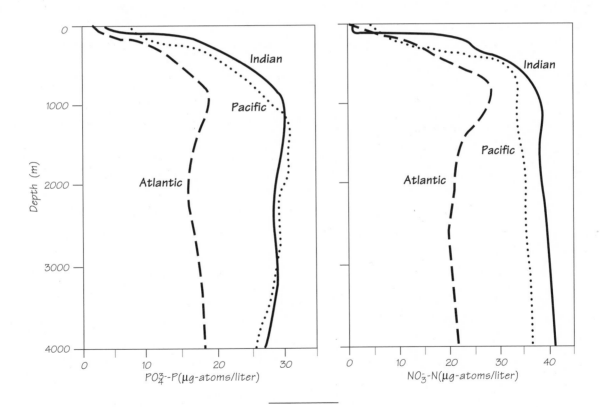

Vertical distribution of phosphate and nitrate
in the three major oceans.

PLANTS AND ANIMALS

equator and the waters surrounding Antarctica also benefit from offshore upwelling. Other highly productive zones are associated with shallow coastal waters, especially where they are enriched by continental nutrient runoff. This enrichment, leading to highly unbalanced N:P ratios, is especially high in estuaries receiving large volumes of sewage and fertilizer runoff.

Measured differences in the productivity of oligotrophic waters of subtropical seas and eutrophic waters of upwelling and coastal zones extend commonly over an order of magnitude, from annual means of less than 50 g C/m² to values approaching 1 g C/m². Global estimates of annual *phytoplankton* production range between 80–100 billion dry tonnes, roughly two-thirds to four-fifths of the terrestrial phytomass total. In contrast, *phytoplankton*'s short life span (1–5 days) means that the standing phyto-

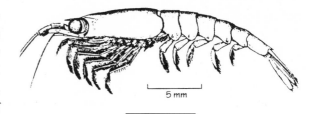

A common species of krill, a shrimplike zooplankton.

mass of the *ocean* is a small fraction of the terrestrial storage.

Spatial distribution of *phytoplankton* shows patchiness on scales ranging from local to global. By far the most revealing grand-scale patterns come from highly accurate estimates of chlorophyll concentrations derived from satellite scanning. They show clearly asymmetric distribution of *phytoplankton* in two nearly concentric bands in the waters surrounding Antarctica. Circumpolar currents and the availability of silicic acid provide the best explanation of this dispersion, while the most intense blooms of *phytoplankton* extend downcurrent from continental masses releasing dissolved nutrients.

Phytoplankton is the energy basis of often highly intricate trophic pyramids. *Food chains* in the *ocean* are commonly complex food webs. A large part of available phytomass is not consumed directly by *herbivores,* but it is first routed through the stores of dead organic matter before bacterial production makes it available to heterotrophs. High phytoplanktonic productivity can support enormous quantities of zoomass. Krill, small crustaceans resembling shrimp and feeding on diatoms, are the most conspicuous zooplankters encountered near the *ocean* surface: their swarms in Antarctic waters can contain up to a billion individuals, and their annual zoomass production total perhaps as much as 1.3 billion tonnes. This prodigious zoomass serves as food for seals, squids, and fish, as well for the biosphere's largest *carnivores,* filter-feeding species of baleen whales.

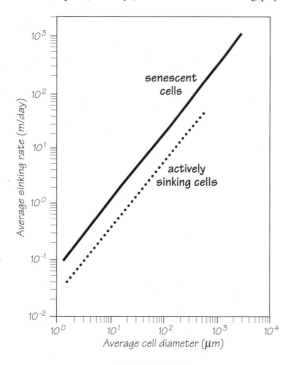

Best-fit lines of sinking rates of phytoplankton.

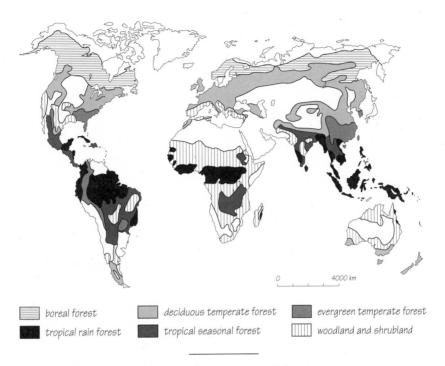

boreal forest   deciduous temperate forest   evergreen temperate forest

tropical rain forest   tropical seasonal forest   woodland and shrubland

0   4000 km

Extent of major forest ecosystems before the
beginning of agriculture.

Forests

Forests store the bulk of the Earth's phytomass, but the share may be as low as three-quarters or as high as nine-tenths. The major reasons for this uncertainty are rapid tropical deforestation, lack of a uniform classification of *forests,* and their high variability. Closed *forests* (as opposed to open woodlands) can be defined as ecosystems whose canopies cover just 20 or up to 40 percent of the ground. Our still poor understanding of tropical *forests* means that we have to extrapolate typical storages from an inadequate number of well-studied sites. The best available inventories put the global area of closed *forests* at about 25 million km² in the early 1990s, with two-fifth of the total in the tropics. The grand total for all *forests* and woodlands was roughly twice as large, and these ecosystems stored about nine-tenths of the planet's phytomass, almost equally partitioned among tropical, temperate, and boreal biomes.

Tropical rain *forests* have the highest average phytomass stores. From the air they project a deceptive sameness of the canopies colored green from low-flying planes, bluish from jets, richly red in false-color satellite images. From the darkened floor, often with only a sparse undergrowth, rise many straight, slender stems, some massive, buttressed trunks, then a jumble of overlapping branches, leaves, lianas and epiphytes; a clearing, or a stream shore, reveals the layered order of the *forest.*

The richest tropical rain *forests* of Amazonia contain nearly 100,000 individual *plants* of more than six hundred species per hectare. But at least three-quarters of stored phytomass (totaling up to six hundred tonnes per hectare) are in only a few hundred canopy and emergent

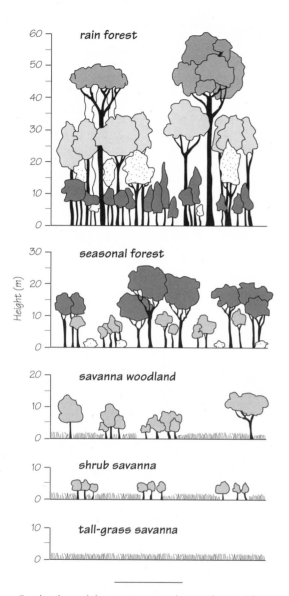

Tree heights and densities in principal types of tropical forests (savanna grasses are shown for a comparison).

trees. Because of the *forest*'s high biodiversity no single tree stores more than a few percent of all phytomass. This evasive strategy has increased survival chances in an environment teeming with seed predators and pathogens. Added active defenses range from smooth barks to symbiosis with guardian ants.

In contrast, temperate and boreal *forests* have just a few dominant tree species, but they can surpass even the richest tropical rain *forests* in total phytomass storage. Unmatched global peaks occur in the Pacific Northwest where coastal redwoods can store up to 3,500 tonnes per hectare of above-ground phytomass, more than five times

The Amazonian rain forest as seen by Henry Bates, a pioneering British explorer, in the 1850s.

Canopy of an Indonesian rain forest.

the total for the central Amazonia. Individual redwoods, above all giant sequoias, are not only the tallest (over a hundred meters) and most voluminous trees, but also the most massive (over three hundred tonnes) life forms on the Earth (the largest blue whales weigh around one hundred tonnes).

Woody above-ground tissues (stems, bark, branches) store most (70–80 percent) phytomass in every *forest;* roots account for 10–35 percent, needles for 1.5–8 percent, and leaves for just 1–2 percent. Merchantable bole, the timber harvested by traditional logging, is only about half of all phytomass: stems with too small a diameter,

stumps, branches, tops, bark, and needles or leaves make up the other half. Typical growing stock of timber in good temperate and boreal stands ranges between 85 and 100 m³/ha (35–50 tonnes per hectare of dry weight depending on the species); in the tropics it can be up to 180 m³/ha. New methods of harvesting for pulp utilize the whole tree (often including also the stump), recovering nearly all of the stand's phytomass.

At about twenty tonnes per hectare, the annual productivity of tropical rain *forests* is nearly twice as high as the growth of temperate and boreal trees. On the other hand, tropical *forests* use available nutrients rather inefficiently. To produce 1 kilogram of new phytomass, tropical trees may require up to twelves grams of **nitrogen,** but boreal conifers may need less than four grams. Fairly high growth rates and economic use of nutrients make temperate *forests* relatively efficient phytomass producers, and high concentration of woody biomass storage in massive stems makes them also easier to manage. Another important advantage for management of these *forests* is their low nutrient-cycling intensity resulting from relatively large litter falls and slow decomposition rates.

As with the standing phytomass, foresters express the annual increments of **wood** in volume, rather than in mass (dry weight equivalents are about 525 kg/m³ for hardwoods and 440 kg/m³ for softwoods). Global means range from 1.4 m³/ha for tropical conifers to 2.5 m³/ha for temperate broadleaves. Managed *forests* have much higher yields: mean for the North American commercial stands is over 2.5 m³/ha, and the fastest annual increments during periods of early growth surpass 10 m³/ha.

All preindustrial civilizations harvested **wood** not only as an indispensable building material but also for fuel, burned directly or converted into **charcoal.** Contribution of *forests* to global primary energy use has been steadily declining with rising **fossil fuel** consumption, but their importance as suppliers of timber and pulp has grown, as has their value as shelters of high biodiversity and providers of irreplaceable ecosystemic services.

Grasslands

The global extent of *grasslands* has changed profoundly since the middle of the nineteenth century. Extensive conversions of North American, Latin American, Australian, and Central Asian *grasslands* to crop fields have greatly diminished natural areas of this biome, but secondary *grasslands* have become more common with advancing tropical deforestation. Although the total area of *grasslands* is almost as large as that of closed *forests,* there is an order of magnitude difference in the average aboveground phytomass storage per unit area (20 t/ha for grasses, 250 t/ha for woody phytomass). But there is more phytomass to *grasslands* than meets the eye: except for the tall tropical grasses, underground phytomass is several times larger than in the canopy shoots.

Two widespread tropical grasses, *Themeda* and *Hyparrhenia*.

Two very common grasses of temperate climates: *Poa* (smooth meadow grass) and *Festuca* (tall fescue).

Live-shoot phytomass ranges from less than one tonne per hectare in semi-desert formations to more than twenty tonnes per hectare in some tropical *grasslands;* inclusion of dead canopy phytomass would raise the maximum aboveground storage values to about thirty-five tonnes per hectare. The greatest live-shoot phytomass is associated with subhumid tropical and maritime climates or with temperate habitats with natural irrigation. General correlation of phytomass with precipitation is unmistakable but the relationship is poor in more humid conditions.

Recorded global extremes of underground phytomass range from less than half a tonne to almost fifty tonnes per hectare, with the lowest accumulation in the tropics, and the highest values in temperate *grasslands* (median close to 20 t/ha). This reflects an expected inverse correlation with temperature, the key determinant of decay rates. The ratio of mean underground to maximum canopy phytomass ranges from 0.2 to just over 10, with the tropical values averaging less than 1.0 and temperate *grasslands* having a mean over 4.0. This lopsided phyto-mass storage is a highly effective adaptation to low, and irregular, precipitation and high temperatures experienced by many *grasslands*. Dense mats of underground phyto-mass act as immense sponges of *rains* and virtually elimi-nate topsoil erosion and provide rich stores of nutrients.

At about ten tonnes per hectare a year, the average productivities of temperate *grasslands* match those of

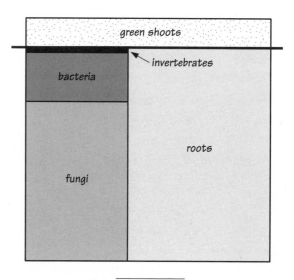

Distribution of biomass in a Canadian prairie grassland: Underground phytomass is much more abundant than aboveground shoots.

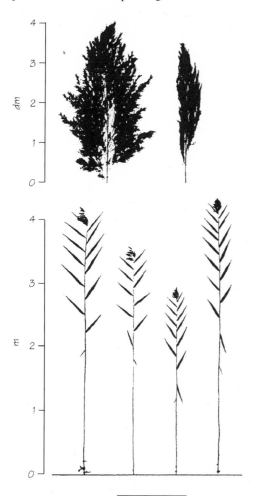

Panicles and shoots of *Phragmites* reeds, common wetland species.

mid-latitude *forests*. Some subhumid tropical *grasslands* can outproduce *forests* of the same region: their maximum annual productivities approach thirty-five tonnes per hect-are—and only in these ecosystems does the share of new production stored in shoots exceed 50 percent of the total phytomass gain. Dependence of productivity on precip-itation is most obvious in the dry tropics: in African *grasslands* every hundred millimeters of *rain* can raise aboveground production by about a tonne per hectare. Wetlands are the most productive ecosystems storing bulk of their phytomass in such dense, and often tall, grasses as rushes and reeds. Reed wetlands occupy at least ten mil-lion hectares worldwide, with the two best known concen-trations in the Danube delta of Romania and the Shatt al-Arab of Iraq.

Grasslands are cropped by a variety of *herbivores*. Only young leaves are high in protein and have relatively high digestibility. Their sheaths, and even more so their stems, are inferior in both respects, but they may account for most of the intake in some species sharing a *grassland* with other *animals.* In the dry season the Serengeti wildebeest get roughly 20 percent of their herbage as leaves and 30 percent as stems, while the respective rates for zebras are less than 1 and more than 50 percent.

A single abundant species of invertebrates can consume a few percent of annual phytomass production, and totals for all invertebrates are typically between 10 and 20 percent. Grazing by ungulates removes up to 60 percent of aboveground productivity in densely stocked East African *grasslands.* Some grasses adapt to high rates of cropping by incorporating digestibility-reducing compounds and toxic substances; other species respond by vigorous compensatory growth following damage done by grazing. In the Serengeti, where grasses are cropped by the world's largest concentration of terrestrial megaherbivores, as well as by many smaller *animals* and numerous invertebrates, moderate grazing stimulates productivity up to twice the level of ungrazed plots.

Lawns offer the most accessible, and often not very welcome, proof of high *grassland* productivity. Grass is now North America's largest crop in terms of area, covering roughly an equivalent of Pennsylvania. Surprisingly, in the humid and temperate eastern part of the continent there may be hardly any difference between the productivity of a lawn cut weekly, fertilized regularly, and watered often, and a lawn neither fertilized nor irrigated and infrequently cut. Although the clippings may seem mountainous, cutting above five centimeters removes as little as 5 and rarely more than 20 percent of total productivity. They may not look it because of their deceptively short stature, but temperate lawns, with typical productivities between ten tonnes per hectare in drier regions and seventeen tonnes per hectare in wetter areas, not only match but also frequently surpass the fixation of many natural *grasslands.*

Heterotrophic Metabolism

Heterotrophs follow two distinct paths in converting the biomass they eat into complex compounds making up their tissues. They can do so through either anaerobic fermentation or aerobic respiration. The first path is restricted to simple prokaryotic cells belonging to fermenting and methanogenic *bacteria,* and to *Ascomycota,* fungi including yeasts that are responsible for the fermentation of *ethanol* (ethyl alcohol).

The second mode became possible only after *plants* increased the atmosphere's oxygen content to the point where some prokaryotes could use aerobic respiration to generate *adenosine triphosphate* more efficiently than by fermentation. Oxidation has clear energetic advantages: lactic acid fermentation liberates 197 kJ for each mole of glucose, alcoholic fermentation yields 232 kJ — but a complete oxidation of that sugar releases 2.87 MJ, a twelve- to fourteen-fold gain. *Nutrients* for the *heterotrophic metabolism* must come from the digestion of plant tissues or from eating other heterotrophs. *Heterotrophic metabolism* has a number of remarkable regularities at organismic

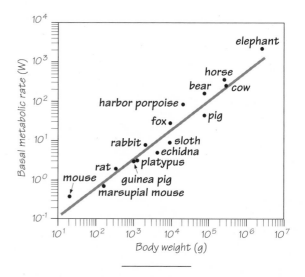

Kleiber's mouse–elephant line of basal mammalian metabolism.

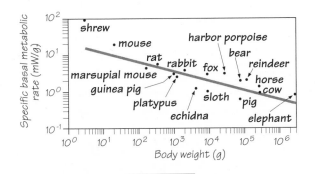

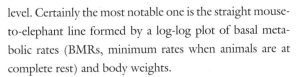

Declining specific metabolism of mammals whose body masses span nearly six orders of magnitude.

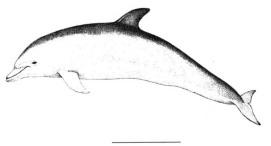

The bottlenose dolphin exemplifies marine mammals with BMRs significantly above the rate expected for their mass.

level. Certainly the most notable one is the straight mouse-to-elephant line formed by a log-log plot of basal metabolic rates (BMRs, minimum rates when animals are at complete rest) and body weights.

The line was first described by Kleiber in 1932 and it showed that animal BMRs (in W) are related to body weight (w, in kg) as $3.52w^{0.74}$. Kleiber's exponent differed from the traditionally assumed value of 0.67, which would be valid if BMRs were a direct function of body surface area. Subsequent additions of BMRs for hundreds of species confirmed the new slope, and in 1961 Kleiber proposed a rounded expressions of $3.4w^{0.75}$ (in W).

Definite explanation of this ¾ power law has been elusive. Analysis of mechanical requirements of animal bodies offers one key. Elastic criteria dictate a proportional relationship between the cube of the critical breaking length (l) and the square of the diameter (d) of loaded animal bones whose diameter is proportional to $w^{3/8}$. Muscle power depends only on their cross-sectional area (proportional to d^2) and hence the maximum power is related to $(w^{3/8})^2$ or to $w^{0.75}$. An even more fundamental explanation rests on the geometry and physics of a network of tubes needed to distribute resources and remove wastes in animal bodies. These space-filling fractal networks

The three-toed sloth is a sluggish metabolizer.

dictate both structural and functional properties of cardio-vascular and respiratory systems, and their properties require that the *metabolism* of entire organisms scales to the ¾ power of their mass.

While the lines of BMRs plotted separately for various animal groups conform to the general slope, their positions differ substantially from values predicted by the general equation. For example, the constant for marsupial mammals (2.33) is 30 percent lower than for eutherian *animals,* whereas the multiplier for songbirds is at least 30 percent above the value for other flyers. Metabolic slopes for invertebrate groups range widely, from less than 0.67 to more than 1.0, but enough of them cluster around the 0.75 line to conclude that that slope is also a valid general approximation for cold-blooded heterotrophs. But, naturally, BMRs of *ectotherms* are only a fraction of those of equally massive *endotherms.*

The Kleiberian exponent has an important obverse in declining specific BMRs (that is total BMR divided by

Fractal geometry of the vertebrate vascular system helps explain the variation of an organism's metabolic rate in proportion to the ¾ power of its mass.

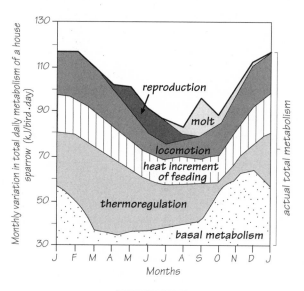

Annual march of total metabolism for an American domestic sparrow.

body mass). This relationship restricts the size of the smallest *endotherms,* but it makes it much easier for larger creatures to cope with prolonged environmental stresses. The daily nectar intake of small hummingbirds is equal to half of their body mass (for people food consumption is just around 3 percent of their body mass), and *endotherms* smaller than these tiny birds would have to feed constantly in order to compensate for rapid heat losses. In contrast, low BMRs of large mammals make it possible for them to go for days without any food and to draw on accumulated fat reserves for relatively long periods of reduced *metabolism* during hibernation.

Departures from the principal trend illustrate various modes of environmental adaptation. In order to thermoregulate their bodies in cold water, seals and whales have BMRs about twice as high as other *animals* of their size, while low specific BMRs of desert mammals reflect their adaptation to periodic food shortages and to recurrent or chronic scarcity of water.

Naturally, BMR accounts for only a part of total energy needs. Digestion raises metabolic rates in all *animals,* and *reproduction* makes periodic energy claims (as do the changes of plumage and pelage in birds and mammals). Search for food is an ongoing activity for all nonhibernating *animals.* Even just standing requires about 15 percent above the resting rate in birds and up to 25 percent in mammals (except in *horses*). Metabolic scopes, multiples of BMR during maximum effort, are naturally highest during *swimming, running,* and *flying.*

Reproduction

Modes of heterotrophic *reproduction* span a continuum between the extremes of early mass breeding in a single spasm and delayed spaced births of a single neonate. The first extreme maximizes the production of rapidly maturing individuals, and life knows no greater opportunists

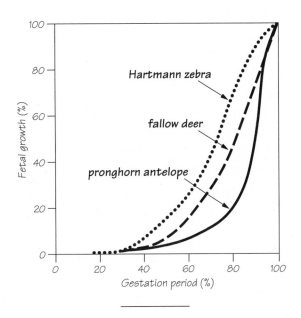

Relative fetal growth (as percentage of birth weight) in three ungulate species.

	r - selection	K - selection
reproduction	early	delayed
body size	small	larger
population size	variable in time, usually well below carrying capacity of environment	fairly constant in time, equilibrium; at or near carrying capacity of the environment
intra- and interspecific competition	variable, commonly lax	usually intense
mortality	often catastrophic, density independent	more directed, density dependent
length of life	short	longer
outcome	productivity	efficiency

Principal correlates of r- and K-selection.

than these r-selection species. Most *bacteria* are such aggressive reproductive opportunists, and so are many species of insects: in suitable conditions they invest so much of their *metabolism* into *reproduction* that they can become objectionable pests. In a few summer days tiny insects such as aphids can pour as much as 80 percent of their metabolic production into *reproduction.* This strategy greatly shortens the parental survival and reduces possibilities of repeated *reproduction.* Endoparasites are an unfortunate exception to this restriction: effortless supply of abundant energy means that tapeworms can reproduce copiously and can survive for up to fifteen years!

On the other end of the reproductive range are K-selection species with much delayed but repeated breeding of very few individuals that take a considerable time to mature. Resulting low rates of increase and poor colonizing ability of these species are more than compensated for by their longevity, competitivity, and adaptability, and by their often highly developed social behavior. Regardless of the position along the reproductive spectrum, fundamental commonalities of biochemical transformations underlying the production of gametes and the *growth* of embryos make it possible to generalize about conversion efficiencies of heterotrophic *reproduction*. Theoretical maximum for organizing food-derived monomers into biomass polymers is about 96 percent, an impressive figure by any standard. Inevitable inefficiencies during digestion of nutrients and production and turnover of tissues will take their cumulative toll, but rates just over 70 percent could be achievable.

Actual performances are easily measured in rapidly reproducing unicellular heterotrophs: they are highest for *bacteria* (50–65 percent), and they average between 40 and 50 percent among yeasts and protozoa. Not surprisingly, sedentary *ectotherms* are the most efficient convertors of nutrients to new zoomass among higher

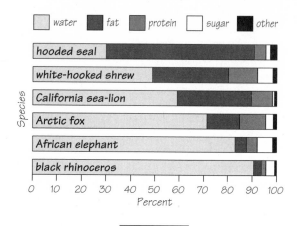

Composition of some mammalian milks.

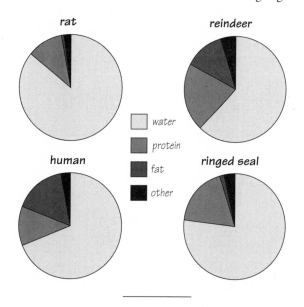

Average composition and total (wet weight) energy content of some mammalian full-term neonates.

heterotrophs: their rates often approach the best possible efficiencies of 70–80 percent.

Among vertebrates, *endotherms* have much faster rates of fetal *growth* than ectothermic species. Thanks to the prominence of the egg in avian life, ornithologists were among the earliest students of the energetics of *reproduction*. Energy needed for testicular *growth* in birds is just 0.4 to 2 percent of basal *metabolism* during the period of rapid gonadal gain. Female gonadal *growth* claims commonly shares three times as large as in males, but this is still only a small expenditure compared to the cost of producing and incubating eggs.

Energy density of eggs is clearly related to the development of hatchlings: precocial species, above all waterfowl, have heavier yolks and shells and their eggs have up to 7.6 kJ/g; in contrast, small altricial birds (emerging from the egg blind and featherless) have eggs containing as little as 4.1 kJ/g. But the extreme case of high energy investment in an egg is New Zealand's kiwi. Eggs of this flightless bird are large (at 400–435 g, roughly a fifth of the female's weight), contain 10 kJ/g, and provide nutrition for one of the longest incubation periods (70–74 days) resulting in a fully feathered chick with an adult-type plumage. Energy efficiency of embryonic *growth* inside an avian egg is about 60 percent: incubation of a chicken egg containing 360 kJ produces a chick whose tissues add up to approximately 160 kJ.

In mammals the cost of gonadal *growth* is negligible compared to the investment in the gravid uterus, growing embryo, and enlarged mammary glands. Their gestation time increases with higher adult body mass, from just twenty days for the European shrew to around a hundred days for the largest *carnivores,* commonly twice as much among large *herbivores* and up to 660 days for African elephants. Intrauterine *growth* rates are fairly uniform in many species (including primates), but they may show dramatic spurts just before the birth (pigs are in this category). Maintenance needs of the gravid uterus dominate the costs of *reproduction* in all mammals except for the smallest rodents and insectivores, whose very short gestations and large litters (commonly five to seven neonates) make the embryonal *growth* more demanding. Weight of newborns as a function of maternal mass generally de-

The kiwi invests an exceptional amount of its energy into an unusually large egg.

creases with size, from about 40 percent for a mouse litter to less than 2 percent for a baleen whale.

That mammalian females often eat in excess of their gestation needs is easily understandable; the energy cost of lactation is considerably higher than that of the gravidity and the females are accumulating fat reserves for the future conversion to *milk* needed for neonate *growth.*

Growth

Neonate heterotrophs differ greatly in the amount of growing they have to do to reach adult mass. Some bats have only to treble their birth weight; primates have to increase it ten to thirty times, pigs two hundred times, and American black bears more than three hundred times. Among vertebrates fish and amphibian neonates fend for themselves, while birds feed their young with regurgitated biomass. Mammals digest a surprising variety of *milks:* dry matter in the liquid ranges from less than 9 percent in asses to extremely rich 64 percent in northern elephant seal.

In general, the most dilute *milks* are those of ungulates and primates, the most concentrated those of aquatic

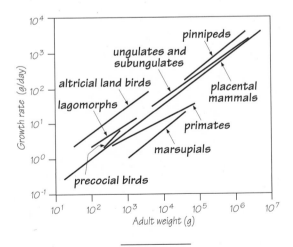

Growth rates of various vertebrates.

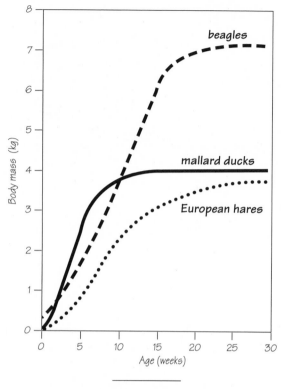

Average growth curves for beagles, mallards,
and European hares.

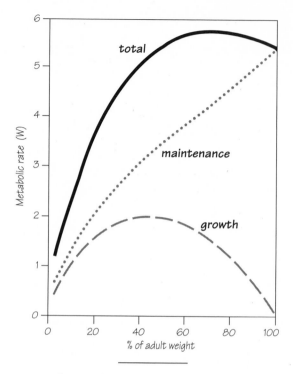

Changing shares of energy required for growth
and maintenance of the white Leghorn chicken.

mammals who must counteract high conductive heat loss in cold waters and deposit rapidly the insulating layer of subcutaneous fat. Seal *milk* has up to 21 kJ/g, ungulate *milks* as little as 1.45 kJ/g. Cow's milk rates 2.7 kJ/g, human *milk* averages 3.2 kJ/g, a closeness of match with obvious importance for the domestication of *cattle*.

Growth rates of nursing and weaned neonates have been fitted into a variety of sigmoid curves with most of the gain coming during a fairly linear intermediate stage when the maximum gains (g/day) can be approximated as power functions of adult weight (w, expressed in grams): the relationship for placental mammals is $0.0326w^{0.75}$.

Notable departures from this trend are the fast-growing aquatic *carnivores* and the slow-growing primates. Energy content of these gains ranges as a function of different concentrations of body constituents.

Mammalian neonates average about 12 percent of protein and 2 percent fat, but grey seals are 9, guinea pigs 10, and humans 16 percent fat. Consequently, energy density of humans averages 8.75 kJ/g at birth, compared to typical mammalian values of just over 3 kJ/g. Energy density of mammals frequently doubles by maturity.

Neonate *growth* cannot be as efficient as embryonic development. Not surprisingly, sedentary ectothermic *animals* are the most efficient convertors of nutrients to new zoomass, often approaching the best possible efficiencies

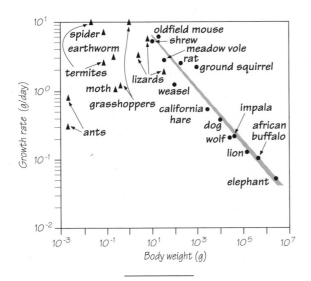

Relationships between production/biomass ratios and adult body weight for a variety of invertebrates (disorderly) and mammals (fairly predictable).

of between 70 and 80 percent. Conversion efficiency of mammalian gains is remarkably similar at the earliest *growth* stage: the average postnatal *growth* efficiency is about 35 percent.

Humans are a great exception: infants deposit only about 5 percent of their food intake as new biomass while doubling their birth weight. Heterotrophic conversion efficiencies decline exponentially with age, to about 15–20 percent when reaching a quarter of adult weight and to well below 10 percent when nearly mature. Similarly, avian *growth* efficiencies decline from around 30 percent in the earliest period of *growth* to just around 5 percent when the birds reach nine-tenths of adult weight.

An interesting evolutionary trade-off works toward equalizing conversion efficiencies of different mammals. *Herbivores* have low absorption efficiency, an understandable consequence of consuming such difficult-to-digest polymers as cellulose and lignin, but high net conversion rates. *Carnivores* have high absorption rates but rather low net *growth* efficiencies, a result of a much more mobile way of life. Consequently, lifetime conversion efficiencies of these two groups are quite similar.

Ectotherms and Endotherms

Thermal limits for all macroscopic life are fairly narrow: proteins start denaturing as their temperature rises above 45°C, intracellular water freezes and the ice crystals tear the cells below 0°C. So how do the arthropods and vertebrates survive in environments ranging from windswept polar ice plains with winter minima below −60°C, to subtropical deserts with daily maxima above 50°C? No less importantly, how do heterotrophs cope with daily temperature fluctuations which may exceed 30°C, and which range commonly between 10–15°C?

Ectotherms and endotherms provide two very different, but both very successful, evolutionary answers to this thermal challenge. All *ectotherms*—that is all arthropods, fishes, amphibians, and reptiles—regulate their body temperature by selecting desirable microenvironments. Salamanders are satisfied with just 10°C, desert lizards prefer 35–40°C. Because of their low *metabolism* and poor body insulation, body size is a critical determinant of sustainable body temperatures in all terrestrial *ectotherms*. Large *ectotherms* can maintain their maximum body temperature ranges within a narrow range, and their considerable thermal inertia allows for longer periods of daily activity.

Tiny insects cannot raise their temperature above the ambient level, but they can change it easily by moving around. Some *flying* insects can actually be endothermic during brief periods before take-offs when they warm up their flight muscles by shivering. Winter moths can reach thoracic temperatures of 30°C even at near-freezing temperatures, but this feat demands so much energy that they

Ectotherms		Endotherms
10^0 - 10^5	body mass (g)	10^1 - 10^7
0.25 - 0.30	standard metabolism (mW/g)	1 - 10
2 - 10	metabolic scopes	6 - 30
20 - 40	growth efficiencies (%)	1 - 4
10 - 80	maximum predator prey ratios (%)	1 - 3
0.1 - 1	daily movement distances (km)	0.5 - 12
weak	social behavior	developed

Principal organismic differences resulting from
the two different thermoregulative strategies.

must spent at least 99 percent of time inactive under insulating layers of leaves and snow.

Notothenioids, Antarctic bony fishes resembling perches, survive in the world's coldest waters by synthesizing at least eight different glycopeptids. When these antifreeze compounds are adsorbed to minute ice crystals they inhibit their growth in body tissues, and the fish can swim in ice-laden waters at –2°C. Antifreeze proteins are also used — together with ice-nucleating proteins initiating the formation of tiny extracellular ice crystals — by some turtles, amphibians, and caterpillars that can freeze solid while preserving their vital intracellular structures.

Their vast numbers and a great variety of niches prove that *ectotherms* have been a clear evolutionary success. But the constant portable microenvironment of *endotherms* has given them a strong competitive edge in benign ecosystems and it enabled them to radiate to even the most inhospitable parts of the biosphere. Endothermy, the maintenance of core temperatures of 36–40°C in most mammals and 38–42°C in birds, is an evolutionary compromise balancing the benefits of a stable core temperature

maintained near the biochemical optimum with the risks of heat death and with high energy costs of *metabolism* and insulation.

Thermoregulation at less than 30°C would need less energy but much higher evaporative cooling. With surface body temperatures often lower than those of the surrounding air, there would be no conduction or convection losses, but high evaporation rates would bring higher risks of dehydration, restricting the diffusion of *endotherms* in arid environments. Higher evaporative cooling would not work with heavy fur insulation, and this would restrict en-

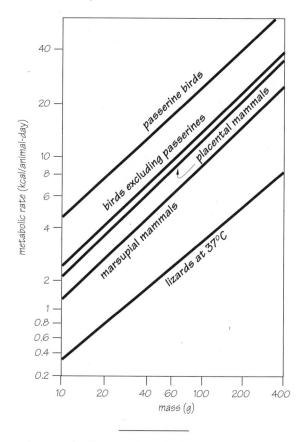

An order of magnitude gap between the metabolic
extremes: Small songbirds metabolize more than
ten times faster than equally light lizards.

Limits of endothermy: life-size image of
a ruby-throated hummingbird.

dothermic diffusion in cold habitats. And lower core temperatures would reduce the efficiency of biochemical reactions.

Small *endotherms* are most disadvantaged because of their higher convection losses, and because of their need for a rapid metabolic response to falling temperatures. A single cold August day may be deadly for a 10 gram hummingbird, while no hibernating animal heavier than about 5 kg has to reduce its winter body temperature by more than a few degrees centigrade. Feathers and furs are outstanding insulators: even at −30°C Arctic mammals have skin temperatures comparable to those of a well-clothed man.

Aquatic *endotherms* have an especially hard task because water conducts heat more than twenty times faster than the air. Even a thick layer of blubber is not enough for the smallest aquatic mammals: their basal *metabolism* must be two to three times higher than the rates of similarly sized terrestrial mammals, and hence they must also find and digest up to three times as much food.

Coping with heat is a different, but not an easier, challenge. A desert *endotherm* with basal metabolism of 50 W/m² may move over sand receiving more than 1000 W/m² of *solar radiation.* Sweating and panting are the best active responses, and both require careful management of water balance. Many small desert *animals* thrive on air-dry food without any liquid water, void concentrated urine and feces, and hide in burrows. Camels forage on dry *plants,* tolerate long spells without water, can lose up to 40 percent of their body mass, and can rehydrate rapidly by drinking an equivalent of more than 30 percent of body weight within ten to twenty minutes.

Food Chains

Rapidly diminishing transfer of biomass energy to successive feeding (trophic) levels is perhaps the most obvious demonstrations of thermodynamic imperatives in heterotrophic lives. Many ecologists would also consider it the central theory of their science, as important as the theory

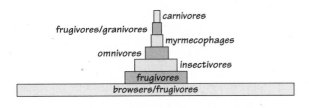

Biomass pyramid of mammalian fauna in an Amazonian forest.

of evolution. Curiously, systematic studies of *food chain* dynamics began only in the early 1940s with Hutchinson's tracing of of zoomass productivity, and with Lindeman's pioneering analysis introducing the notion of energy transfers and efficiencies. Subsequent investigations revealed a number of surprising realities about these fundamental ecosystemic relationships.

Numbers of *herbivores* are most often limited by the presence of *carnivores* rather than by the availability of phytomass; this means that grazers and leaf- and seed-eaters are rarely in direct competition. In contrast, numbers of *plants* (primary producers), decomposers, and *carnivores* are limited by the availability of their respective resources, a situation leading inevitably to considerable interspecific competition. *Food chain* length increases along a primary productivity gradient. Diminishing transfers of energy to higher trophic levels obviously limit the length of *food chains*—but in many instances the *chains* are truncated well before this energetic limit is reached in order to minimize feeding risks in fluctuating environments.

Food chains average just three links in terrestrial ecosystems. Ocean ecosystems based on primary production by *bacteria* and *phytoplankton* have particularly complex food webs. Five trophic levels are usual in aquatic environments, kelp beds have six levels, and some coral reefs have seven levels. Actual transfer efficiencies vary greatly among different ecosystems and trophic levels.

Herbivores consume less than 10 percent of available phytomass in temperate *forests* and fields and up to 60 per-

cent in some tropical *grasslands.* In every ecosystem a major share of phytomass is consumed by soil fauna and by insects; aboveground vertebrates may harvest as little as 1 percent of net primary productivity in some temperate environments. Ingested phytomass is converted to herbivore zoomass with efficiencies ranging typically from 40 to 60 percent for *ectotherms* and 60 to 90 percent for *endotherms.* Consequently, efficiencies of energy transfer at the primary consumer level fall mostly between 1 and 15 percent, departing significantly from the commonly assumed mean of 10 percent.

Carnivorous *ectotherms* convert as much as 30–35 percent of ingested zoomass into their body tissues, but the rate for their endothermic counterparts is only 2–3 percent. Low efficiency of energy transfer is a price mam-

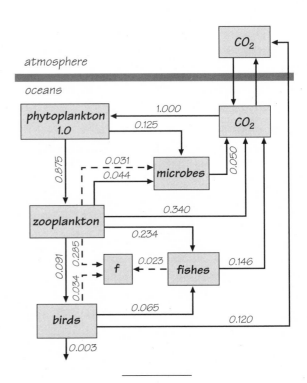

Carbon allocation in a typical Antarctic marine food web.

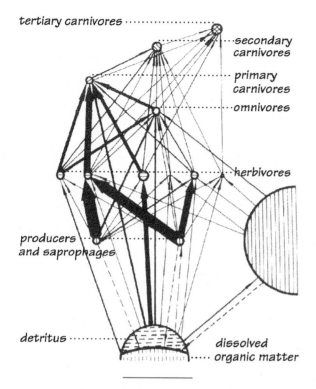

tertiary carnivores

secondary carnivores

primary carnivores

omnivores

herbivores

producers and saprophages

detritus

dissolved organic matter

Planktonic food web in the Black Sea.

mals and birds pay for their high rates of *metabolism* and for their highly mobile way of life. This combination also limits their total standing biomass to a minuscule fraction of the planetary total dominated by *bacteria*, microscopic fungi, and soil invertebrates. These largely invisible heterotrophs are the evolutionary basis of the trophic pyramid: they can thrive without vertebrates, but all higher heterotrophs depend on their irreplaceable roles as decomposers and recyclers of nutrients.

While the fraction of energy transfered to higher trophic levels declines to a small share of the initial biomass, there is no corresponding regression of heterotrophic sizes and weights. In even-linked *grasslands* (two trophic levels) the dominant grazers may be ungulates (wildebeest) much larger than their predators (cheetahs).

In odd-linked tropical *forests* (three trophic levels) small *herbivores* (monkeys) and insectivores are controlled by larger predators (eagles and jaguars). In the *ocean* the most abundant herbivorous zooplankters are larger than autotrophs (*bacteria* and *phytoplankton*), their consumers (fish, squid) larger again, and predators of predators (tuna) still larger, forming an inverted biomass pyramid.

No precise quantifications of the global trophic pyramid are possible, and we cannot be even certain that all of the best estimates capture at least the right orders of magnitude: rapid turnovers of decomposers and invertebrates make their zoomass estimate especially difficult. Above-ground phytomass, the pyramid's foundation produced by *photosynthesis,* stores around one trillion tonnes of dry matter equivalent to about 2×10^{22} J.

Grand total of terrestrial heterotrophic biomass appears to be around eight billion tons of dry matter, or slightly less than one percent of all phytomass stores. Decomposers account for at least four-fifths of the heterotrophic total, invertebrates for roughly a tenth, and vertebrates contributes less than a twentieth. Zoomass of domestic *animals,* dominated by large ungulates, is now perhaps as much as twenty times the total of all wild vertebrates, and the mass of our species, now approaching 100 million tonnes, is an order of magnitude larger than the total for all non-domesticated mammals.

Herbivores

Energy imperatives dictate that *herbivores* must be the most abundant heterotrophs — but they do not determine their individual size. Densities of heterotrophs decrease exponentially with their body weight: the complete plot ranging from the smallest *bacteria* to the largest mammals has a slope of −0.75, which means that the exponent for density decrease with body length will be −2.25. On the average, every square kilometer of land will support only few hundred vertebrates (reptiles, birds, mammals) with

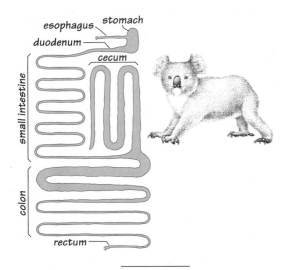

Adaptation to extreme herbivory: Cellulose in eucalyptus leaves, the koala's only diet, is digested in an unusually long cecum by microorganisms.

body lengths of 20–50 cm — but about 10 million invertebrates measuring 0.2–0.5 cm.

And because the rate of basal *heterotrophic metabolism* increases with the 3/4 power of the body mass, energy harvested daily per unit area should be a constant independent of the size of individual *herbivores.* This means that, as a general rule, no herbivorous species has a competitive advantage just because of its bigger size. Bigger bodies mean that although there are fewer individuals of the same species (and also fewer species and fewer species per genus) they intercept a share of biomass energy roughly identical to that eaten by vastly more numerous tiny heterotrophs. Indeed, in areas where they coexist, ants and elephants consume annually very similar amounts of phytomass, between 200–350 kJ/m².

Not surprisingly, there are many departures from this general relationship. Similarly sized *herbivores* may be

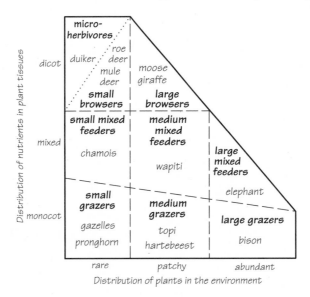

Feeding niches of ungulates.

lateral view of skull   **ventral view of upper dentition**

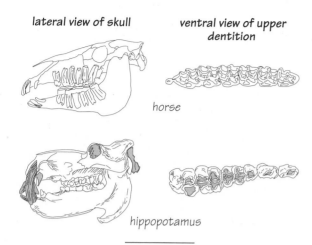

horse

hippopotamus

Skulls and molars of two large herbivores.

Herbivores feeding on the least energy-dense substrates—be they silica-reinforced grasses or toxin-infused leaves—require effective adaptations. Large grazers have wide jaws and incisors suitable for high-volume grass mowing. To facilitate digestion, they have large rumens for prolonged fermentation of fibrous phytomass. Megaherbivores, grazers weighing more than one tonne, offer the best illustration of this adaptive success. They can tolerate substantial fluctuations in feed availability, they can thrive on a poor-quality forage and yet attain high zoomass densities (elephants up to 5, buffaloes over 3 g/m^2), and their huge size makes them invulnerable to all natural predators. Megaherbivores used to be more diverse and inhabit all continents, but human predation eliminated mammoths and has come close to eliminating the only surviving species—elephants, rhinos, hippos, and giraffes.

The ultimate mechanical adaptation to grazing is found in manatees, aquatic megaherbivores (maxima over 1.5 tonnes) living in both fresh and ocean waters of Florida, the Caribbean, Amazonia, and West Africa. Manatees are continuously sprouting new teeth at the rear of their jaws that move forward as the frontal teeth, worn out by abrasive freshwater grasses, drop out. Small leaf-eaters are no less adept at coping with poor diets. Koalas, feeding exclusively on leaves of various eucalyptuses, do not only rely on microorganisms residing in elongated cecum, an

present in densities differing by more than an order of magnitude, with migration making frequently most of the difference. Serengeti's nomadic wildebeest, whose adults average around 200 kilograms per head, follow rains, and various estimates put their total between fifteen and thirty-five times that of nonmigratory topi weighing about 90 kilograms per animal.

And in unusually productive ecosystems, above all on Africa's tropical *grasslands,* zoomass of the largest mammalian grazers could actually surpass that of common soil invertebrates. During the 1960s, before the slaughter of Uganda's protracted civil wars, the Ruwenzori National Park supported as much as two hundred kilograms of large *herbivores* per hectare, and parts of the Serengeti still carry more than one hundred kilograms per hectare—while the densities of tropical termites range mostly between ten and fifty kilograms per hectare. In contrast, total zoomass of highly specialized *herbivores* that can digest only a small part of available phytomass in widely dispersed seeds, fruits, and nectars, is very limited: even in the richest tropical rain *forests* live weight densities of avian faunas usually do not surpass 0.1 g/m^2.

Messor ants survive in very dry environments by feeding exclusively on seeds.

expansion of the hindgut at the junction of the two intestines, to ferment the leaves, but they can also neutralize toxic oils present in the foliage.

Approximate equality of energy harvests and great disparities in individual body size translate into major differences in the energy flux through herbivorous communities. Ratios of annual phytomass consumption to standing zoomass are less than ten for elephants but as high as eight hundred for the smallest rodents. These tiny mammals also excel in producing new zoomass: production/zoomass ratios for shrews and mice are mostly between two to three, compared to mere 0.05 for elephants.

Neither of these production extremes — rapid *growth* of small *animals* or slow accumulation in huge bodies — would be suitable in *herbivores* domesticated for *meat* production. Ideal meat animals should combine relatively fast production with substantial zoomass accumulation. Pigs meet best these two requirements: modern meat breeds eating high-protein feeds (mostly corn and soy-

beans) can go from neonates of few hundred grams to a hundred kilograms in just six months.

Carnivores

As with *herbivores,* most carnivorous species are not large, charismatic mammals. By far the most abundant predators are inconspicuous insectivores, including thousands of insect species using unusual hunting techniques — ranging from swarming raids by army ants to net and poison immobilization by spiders — to capture prey of the same, or even larger size than themselves: raiding ants can kill even lizards, snakes and nestling birds. And the vertebrates who grow to the largest sizes by eating other heterotrophs are not bears and lions but marine suspension-feeders.

Basking sharks are continuous ram feeders *swimming* with their mouths open. Baleen whales are intermittent ram feeders rushing forward to take enormous gulps of water. They use screens of finely fringed comblike plates hanging from their upper jaws to strain masses of krill,

Detail from Peter Brueghel the Elder's
Big Fish Eat Little Fish (1556).

In just a few milliseconds a frogfish's mouth can extend enormously to suck in the victim lured within its reach.

tiny crustaceans feeding on *phytoplankton,* as well as small fish. The largest baleen species, finback (rorqual) and blue (sulphur-bottom) whales, can grow to lengths of between twenty and thirty meters and weights of fifty to a hundred tonnes.

For most of the smaller swimmers, and for birds and terrestrial vertebrates carnivory is not so easy, and they have adopted a continuum of hunting strategies suited to their mobility. Vertebrate *ectotherms,* such as reptiles and amphibians, cannot rely on pursuit, and must spend much time waiting in ambush and then launching rapid strikes. But the ultimate ambushers are tropical frogfishes. They can match the color of the substrate on which they rest motionless, resembling rocks.

Their elongated dorsal-fin spine, extending upward between their eyes, is topped with a lure that looks like a small fish or a crustacean. A rockfish wiggles the lure when a prey approaches, and then it pounces and swallows it whole faster than any other vertebrate predator. And it can

do so easily even when the prey is much bigger than itself: in a split second its mouth can expand twelvefold!

Many fishes, lizards, and birds feeding on small invertebrates use saltatory search, where pauses to look for prey alternate with advances. At the other extreme of the hunting continuum are some large oceanic *carnivores*—tuna, swordfish, and most sharks—and large birds of prey that are almost constantly on a cruising search. Yet even the most successful hunters cannot attain great prominence in ecosystems. Inevitable losses of useful energy along *food chains* limit both the size and the number of *carnivores.* The total zoomass of all predators in the Ngorongoro Crater, perhaps the Earth's best remaining place for predators to hunt large ungulates, is less than 100 kg/km² (about 0.7 kJ/m²), merely one percent of the crater's total density of *herbivores.* Live weight of cheetahs in the Serengeti, one of their last extensive natural habitats, is equal to just 0.1 percent of the total zoomass of ungulate grazers.

But carnivorous existence has its obvious rewards. Even smaller vertebrate *carnivores* are rarely preyed upon

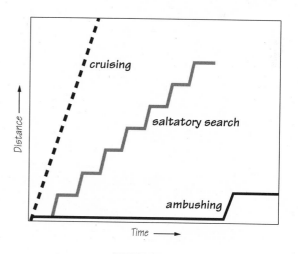

Principal hunting strategies of carnivores: ambushing, saltatory search, and cruising.

(but cannibalism of neonates is not uncommon even among the largest species, including lions), and larger ones have no natural enemies. Low densities of larger **carnivores** mean that they are rarely in direct competition. While many **herbivores** have to resort to specialized mass-processing of low-quality phytomass, predators feed on a much more easily digestible high-protein zoomass. Consequently, all **carnivores** need only simple digestive systems and yet they can assimilate 85–95 percent of the ingested food. Moreover, in areas with plenty of prey they can spend most of their time resting after gorging on relatively easy kills.

Still, even among the mammals belonging to the order *Carnivora* there are some obligatory **herbivores** (the giant panda is restricted to leaves of several bamboo species) and a number of omnivores (most bears). And some species, including lions, get a good deal of their meat by scavenging rather than by hunting. Canids and felids are the two families of quintessential **carnivores** with distinct structural and behavioral adaptations for hunting smaller and highly mobile vertebrates (mice, rabbits) as well as larger **herbivores** (wildebeest, moose).

Canids — wolves, wild dogs, coyotes, foxes — excel in endurance **running;** they kill by leaping, pinning the prey with their forepaws and then biting it. Felids — lions, leopards, cheetahs, wild cats — capture their prey in fast but short rushes culminating in a leap and a quick killing bite. Success rates of canid and felid hunting can be very low:

only a few percent of solo rushes can end in capture of the largest ungulates. Odds are greatly improved by ambushing the prey (especially near water), and by group pursuit practiced by most canids. The latter technique also allows them to bring down **animals** of much larger size than themselves: wild dogs can kill wildebeest, wolves can subdue a moose.

Swimming

For most fishes and aquatic mammals **swimming** requires little or no energy to support their neutrally buoyant bodies. Their exertions go overwhelmingly into generating the thrust needed to overcome the pressure and friction drag of the relatively dense medium. The product of drag (kgm/s^2) and speed (m/s) predicts the metabolic power (kgm^2/s^3) needed for **swimming.** Because the drag is proportional to the square of speed, power needs of swimmers will increase with the cube of their velocity.

Aquatic heterotrophs have a choice of several distinct manners of propulsion: paddling, jetting, use of hydrofoils, and body undulation. The fastest swimmers are propelled by hydrofoils that are either horizontal (dolphin and whale flukes beating up and down) or vertical (fish tails beating sideways). The front halves of their bodies are inflexible, but there is a rapid increase of wave amplitude over their rears, culminating at the caudal hydrofoil. Recoil movements are prevented by minimizing rate of change of water momentum by a marked reduction of

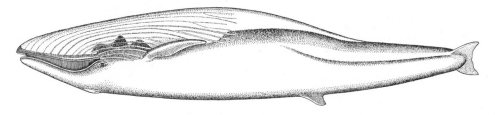

Baleen whales, intermittent ram feeders,
are the world's largest carnivores.

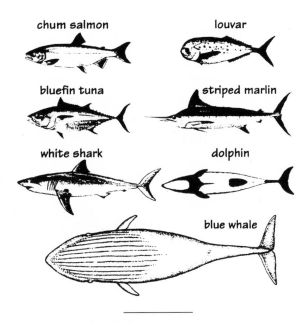

chum salmon
louvar
bluefin tuna
striped marlin
white shark
dolphin
blue whale

Lunate tails of different swimmers (not drawn to scale, dolphin and blue whale shown from below).

cross-section depth just in front of the hydrofoil and by large cross-section in the body center, commonly enhanced by dorsal fins.

As for the hydrofoil itself, all strong swimmers share a remarkably similar lunate tail whose blunt leading edge and fine tapering clearly resemble a well-designed aircraft wing. The maximum measured speed of hydrofoil propulsion is eleven meters per second for a spotted porpoise in Hawaii; the skipjack tuna comes close with ten and a half meters per second. These performances translate to metabolic rates of up to around 50 W/kg of body weight, twice as much as the power output for elite endurance athletes. Some hydrofoil swimmers—dolphins, porpoises, *flying* fish, and penguins—alternate underwater movement with leaping into the air, but lack of proper measurements makes it impossible to conclude if this behavior actually saves energy.

Penguins and sea turtles use a peculiar mode of hydrofoil propulsion: they beat, respectively, their stubby wings and flippers, generating thrust by lift that acts forward and upward on the downstroke (their hydrofoils have a positive angle of attack) and forward and downward on the upstroke (hydrofoils are angled negatively). Ingenious measurements of *swimming* penguins show that they need about 7 W/kg of body weight during aerobic motion averaging about 0.7 m/s (about 2.5 km/h). When they swim at or near the surface the bow wave, which increases with their speed, slows them down and makes them work harder, while *swimming* well submerged may cut their energy expenditure by as much as one half and increase their sustained speed by up to 20 percent.

Undulating swimmers—eels, dogfishes, lungfishes, and sturgeons—move forward by passing backward waves along their bodies, but they can also reverse the waves or use them to move vertically. Continuous dorsal and ventral fins help to maintain the body depth all the way to the tail, improving the thrust by maximizing the mass of water that can be set in motion by anguilliform *swimming.*

Squids move jerkily rearward by squirting water out of their mantle cavity whose capacity limits the speed. Sustainable velocities of this pulse jet propulsion are less than one meter per second for small squids, and they need more

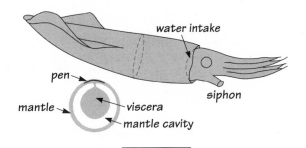

water intake

pen

mantle

viscera

mantle cavity

siphon

Jet-propulsion swimming of squids is done by repeated rapid contraction of the mantle forcing water out through the siphon.

than 1.5 W/kg of body weight. But rapid contractions can more than double that speed, and they are fast enough to propel the cephalopods into fairly high air leaps.

Aquatic birds, paddling with their spread webbed feet which they fold during the recovery stroke, need around 10 W/kg of body weight for sustained *swimming.* They usually move at speeds just below the rate that would generate a significant bow wave, and they cannot improve their efficiency by *swimming* underwater: air underneath their feathers makes them too buoyant and they have to paddle just to stay submerged.

Unlike with terrestrial *animals,* it is difficult to offer general conclusions about locomotion efficiencies and metabolic scopes of fishes because the baseline is so moveable. As *ectotherms* they change their resting *metabolism* by more than an order of magnitude, mainly with the water temperature but also with oxygen content, light, and presence of other fish. Moreover, measurements of *swimming* salmons indicated that their metabolic scope rises with increasing body size, from just four in the smallest fish to over sixteen in mature individuals.

But there is little doubt that for all species well adapted to aquatic life *swimming* is the least energy intensive way of locomotion, and its power needs appear to be

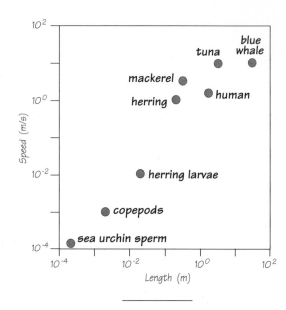

Body lengths and swimming speeds of organisms ranging over five orders of magnitude.

declining with the heavier body mass along a line sloping at approximately –0.3. Unfortunately, our knowledge of actual power needs does not extend to the largest swimmers, to whales weighing as much as one hundred tons. If the declining specific power line held at these high masses, a baleen whale would need less than 100 mJ/gkm and her moving costs would be a small part of her maintenance *metabolism.*

Running and Jumping

Body size imposes fundamental limitations on the energy cost of *running.* Inevitably, smaller creatures must take more steps per unit of distance than the larger ones: the number of steps is inversely proportional to body length, that is approximately to one-third power of the animal's mass. And because the energy spent on each step is proportional to body weight, the cost of *running* per unit distance will be proportional to 0.67 power of the mass,

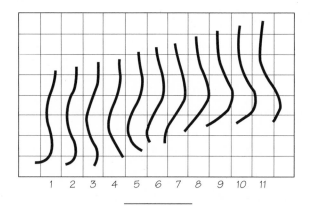

The sequence of anguilliform swimming.

while the cost per unit body mass per unit distance will fall with −0.33 power of the mass.

These economies of scale hold not only for mammals, but also for virtually all other runners, ranging from ants to birds: most regression slopes fit between −0.28 and −0.35, an especially notable congruence given the fact that these diverse species are anything but isometric. This relationship dictates that the cost of *running* at higher speeds will be going up much faster in the smallest animals than in very large mammals. Contrary to earlier assumptions, the work needed to overcome the inertia of limbs is not a substantial factor in the expense of *running.* Animals whose mass is differently distributed over their limbs and bodies still have nearly identical energy costs at different *running* speeds as long as their total body mass is similar: the slope for cheetahs is not radically different from that for goats!

Cheetahs are the fastest-running *carnivores.* Their highest accurately measured speed was 105 kilometers per hour, far faster than *horses* (69 km/h) or greyhounds (58 km/h). But, like any other sprinter, they cannot sustain their speeds for more than a few hundred meters. Studies in Serengeti show that sixty to ninety meters is the usual rushing distance, while the average for successful hunts was merely forty-five meters. Even so, the high energy cost of these rushes—indicated by up to 156 breaths a minute after a dash and kill, compared to sixteen breaths a minute at rest—require subsequent recovery and limit the daily number of such intensive chases.

Some measured runs put Thomson's gazelles slightly ahead of cheetahs, but they too are mere sprinters. In general, because the power available for *running* goes up faster then the cost of the activity—the maximum oxygen intake rises with body mass to the power of 0.85 while the cost of *running* goes up with the power of 0.67—large mammals can run up to ten times faster than the smallest animals. Impressive stampedes of bisons and rapid attacks of bears and hippos clearly demonstrate this capability.

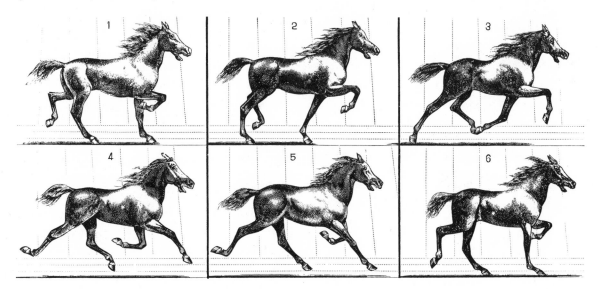

Paragon of elegant running: a galloping horse from Eadweard Muybridge's studies of animal locomotion (1887).

Running is also easier for large mammals because of elastic structures of their legs. Kinetic and potential energy lost at one stage of their strides is temporarily stored as elastic strain in both muscles and tendons to be used later as elastic recoil. At high speeds some large mammals, including men, may save more than half of the metabolic energy they would otherwise need in *running.*

Most of the animal *running* is done in the search for food, but given the diversity of runners there are hardly any simple relationships between foraging and energy expenditure. Daily movement distance (in km) for mammals ranging from rodents weighing sixty grams to six

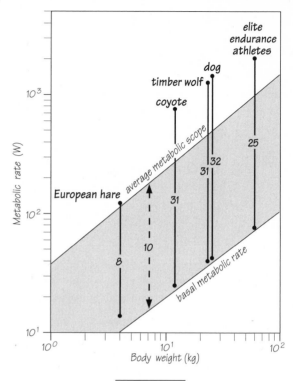

Metabolic scopes of some fast runners.

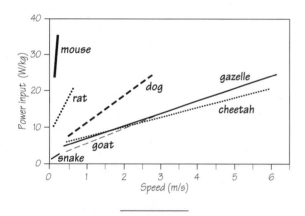

Power input versus speed for different runners.

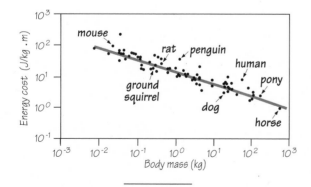

Metabolic energy cost of running for mammals.

thousand-kilogram elephant scales as $1.038w^{0.25}$ (weight in kg) — but it varies by nearly two orders of magnitude for a given body mass. Expectedly, *carnivores* average more than four times longer distance than *herbivores.* Small noncarnivorous mammals may use less than one percent of their total daily energy expenditure on moving around, while large *carnivores* commonly expend 10–15 percent, and frequently even more, of their daily energy flow on locomotion.

Jumping is a particularly interesting case of terrestrial locomotion with a counterintuitive performance limit: regardless of their body size, all animals should jump to roughly the same height. This is a consequence of energy output being directly proportional to the muscle mass, which will be generally proportional to the total body mass: the identical energy release per unit mass should then raise the animals to equal heights.

Actual measurements confirm the similarity of **jumping** performance for animals whose body size differs more than 10^8 fold. A flea (0.49 mg) jumps twenty centimeters, a locust (3 g) sixty centimeters, a man (70 kg) also sixty centimeters (this refers to the lifting of the body center during the best standing jump, which does not utilize the kinetic energy of athletic high jumping). The threefold difference between a flea and a man would virtually disappear if the tiny animal could jump in vacuum: for small insects the air resistance is very important, for large animals it hardly matters.

The exceptions are achieved by disproportionate muscle mass. African bush babies (primates of *Galago* species) can jump up 2.25 meters, but their jumping muscles are about twice as massive as are those of a man. And kangaroos are obviously excellent jumpers but their objec-

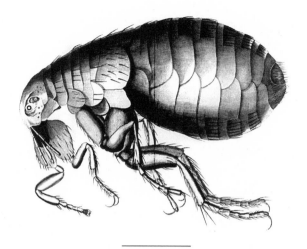

Robert Hooke's flea from *Micrographia* (1665).

tive is distance, not height, as they use their large Achilles and tail tendons to store elastic energy for the next move: their hopping needs less energy than their pentapedal *running*.

Flying

Flying is not uncommon among heterotrophs. Most insects are fairly good, although rather slow, fliers, and *flying* and gliding species can be found in each of five classes of vertebrates. And while there is no shortage of impressive *flying* achievements—ranging from overnight dashes of two hundred kilometers by the oriental armyworm moth in eastern China to echolocating feats of bats—it is the bird flights with its speed, maneuverability, and endurance, which elicits most admiration and wonder.

Birds are perfectly adapted for these accomplishments. Their feathers—light yet strong outer vanes underlain by soft down—enable the airborne maneuverability and provide outstanding insulation. Their forelimbs lengthened into wings are complex aerofoils with downward camber, thickened leading edges, and fine tapering providing sufficient lift and thrust. Large data sets show

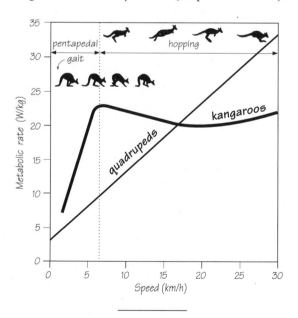

Power input of pentapedal running and hopping for kangaroos.

wing length going up as 0.33 power of the avian mass, while their area increases with 0.67 power.

Internal adaptations are no less remarkable. All birds have large pectoral muscles that can sustain vigorous flapping: regardless of the body size they make up about one-sixth of most bird bodies, and they account for nearly twice as much in tiny hovering hummingbirds. While mammalian breathing is a succession of tidal flows, a complex arrangement of parabronchi and air sacs allows birds to use unidirectional flow through a large number of parallel tubes. Avian parabronchi remove almost 30 percent more oxygen from the respired air than mammalian alveoli. Vertebrate runners become incapacitated at high altitudes, but many birds migrate thousands of kilometers at altitudes between three and six kilometers, and bar-headed

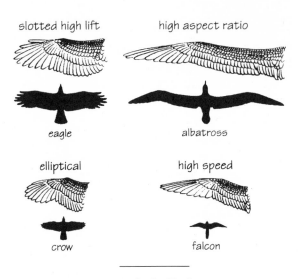

Four major kinds of bird wings.

geese were seen overflying Mount Everest!

Minimum energy costs of *flying* can be easily estimated from the loss of potential energy in gliding. Birds and bats losing 1–2.5 meters of height per second reduce their potential energy (a product of their mass, acceleration of gravity, and height change) by 10–25 watts per kilogram of body mass. Level flight will require at least that much power, but because muscles in prolonged use will not work with more than 20 percent efficiency the metabolic cost of flapping flight should be most commonly between 50 and 130 W/kg.

Experimental measurements of birds and bats in wind tunnels have confirmed these expectations, but they also uncovered two distinct power-speed relationships. While smaller fliers (bats and small Australian parrots) behave as aircraft—they have the lowest specific energy needs at intermediate speeds—larger birds (starlings, gulls, and crows) show no, or hardly any, increase of power needs with increasing flight speeds.

But the largest birds find it very difficult to produce enough power for sustained flight. As their weight in-

A primary wing feather and a permanent fluffy down feather.

creases, the power needed for *flying* rises faster (exponent 1.0) than the power which can be delivered by pectoral muscles (exponent 0.72). Consequently, there are few *flying* birds heavier than ten kilograms and none above sixteen kilograms, the weight of the heaviest Kori bustards in East Africa, which fly only rarely. Not surprisingly, all heavy fliers conserve their energy by spending much of their time gliding rather than flapping.

Some gliders do not even use active flight to gain the initial height: they rely on circular soaring in thermal currents, and they can repeat the feat by gliding from thermal to thermal. European storks use this technique even on their long-distance migration to Africa. And heavy fliers can reduce the energy cost of sustained flapping by *flying* in V-shaped formations. This arrangement saves energy by

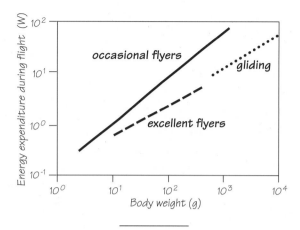

Energy expenditure during flying and gliding.

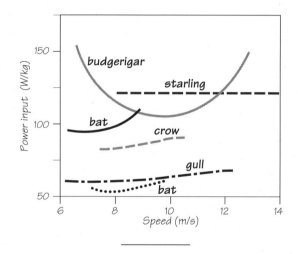

The net metabolic energy cost of flight.

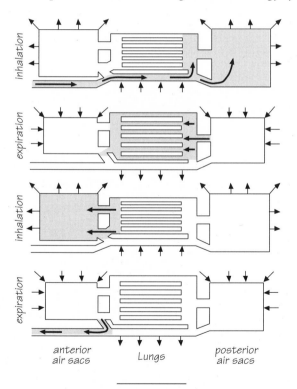

One-way air flow through bird lungs.

leaving vortices only behind the outer wing tips of birds *flying* on the edge of the V. But no matter if in formation or moving alone, heavier birds have to fly faster than small species in order to generate adequate lifts: songbirds average usually less than ten meters per second, Canada geese almost twenty meters per second.

The most admirable *flying* feats—long-distance migrations of insects and tiny birds, including such feats as journeys of North American butterflies to Mexico averaging a hundred kilometers a day and four thousand-kilometer nonstop flights of blackpoll warblers from New England to Venezuela — could not be accomplished without preflight accumulation of *lipids.* Monarch butterflies can deposit up to 125 percent of their lean dry weight in *lipids,* and the ratios of lipid/lean dry mass, which are just around 0.35 in nonmigrating songbirds, can be up to ten times higher in migrating species. *Lipid* conversions can cover all, or most, migratory distances. Birds that fly farther than predicted on the basis of standard flight-cost equations and peak fat reserves are either bit more efficient convertors or are using tailwinds.

For birds lighter than one kilogram, as well as for all insects, *flying* requires less energy than *running* of similarly massive *animals* — but some of the largest birds may need to invest more energy in *flying* than comparably sized mammals in *running.*

3

PEOPLE AND FOOD

As it does in other warm-blooded species, *thermoregulation* keeps the core human body temperature within narrow bounds, but this high degree of similarity does not extend to our metabolic rates. Human diversity, demonstrated in so many fascinating physical and psychical ways, is very evident in surprisingly large individual differences in *basal metabolism*, in energy costs of *pregnancy and lactation* — generally much higher in affluent nations than in poor countries — and in metabolic expenditures on *labor and leisure.*

Balanced intake of essential food *nutrients* is indispensable for normal *human growth,* maintenance of body tissues, and physical and mental activity. Throughout the human evolution most of food energy has come from carbohydrates and, in turn, most of these *nutrients* came from cereal and leguminous *grains.* Cereals have been eaten in a great variety of foodstuffs, but none of them has been of such importance in the Western civilization as *bread.* Rising affluence has transformed typical food intakes: consumption of coarse cereal and leguminous *grains* has declined, demand for processed cereals, sugars,

milk (even in countries with high rates of lactase deficiency), red *meat,* and poultry has risen.

Higher demand for animal foodstuffs has brought not only significant increases in protein intakes, but also in the consumption of *lipids.* As a result, in most rich nations fats and proteins now supply more food energy than do carbohydrates. Higher average consumption of *ethanol* is also generally correlated with rising affluence, but the kinds of consumed alcoholic beverages vary a great deal among different countries.

Basal Metabolism

Determination of an individual's basal metabolic rate (BMR) requires a body at complete rest, in postabsorptive state (the last meal eaten hours ago), and in a thermoneutral environment: most of us would qualify when in our beds at 3 A.M. Physiologists have accumulated thousands of BMR values for both sexes, mostly by measuring gas exchange by indirect calorimetry in respiration chambers (O_2 consumption, CO_2 production).

This huge data base is not unbiased: it includes statistically significant numbers for all ages as well as values for adults of different stature and of different weight for height, but it is made up overwhelmingly of healthy individuals from Western nations. Measurements from poor countries — where inadequate nutrition more commonly prevents people to express fully their physical *growth* potential — are underrepresented.

Body size, composition (share of metabolizing tissues), and age determine complex variations of BMRs. Relationship between BMR and body weight is best captured, for both sexes and for all ages, by simple linear equations. Their fits remain unimproved by inclusion of body surface area or height, but their predictive power is

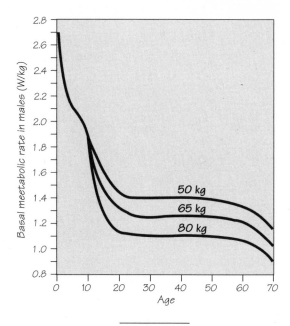

Evolution of specific basal metabolic rates in boys and men.

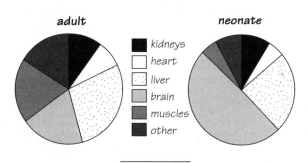

Relative metabolic shares in newborns and in adults.

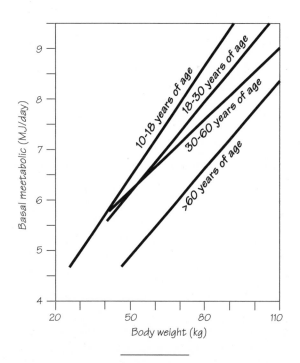

Best fit lines of equations predicting basal metabolic rates for males.

very high only for children and adolescents. Correlations between body weights and BMRs range from impressive 0.97 for children less than three years old to 0.9 for teenage boys. In contrast, adult fits are poor: for men between thirty and sixty, the group forming bulk of the economically active population, the correlation explains only about a third of the variance. In addition, these equations, derived mostly from European observations, tend to overestimate the *basal metabolism* of many non-European populations.

Clearly, BMRs are highly individual, but the key question asked by Elsie Widdowson, an eminent British physiologist, in the late 1940s — "Why can one person live on half the calories of another, and yet remain a perfectly efficient physical machine?" — is yet to be answered satisfactorily. Inevitably, the answers will lie in finding the causes for substantial disparities in the functioning of internal organs which account for most of every individual's *basal metabolism.*

In relative terms, kidneys are metabolically the most active organs, followed by the heart, brain, and the liver. In absolute terms, the liver uses the largest share of BMR in adults (at least one-fifth), the brain in children. A newborn brain, accounting for only a tenth of total body mass, claims typically more than 40 percent of BMR; liver, with less than 5 percent of total body mass, needs a fifth of BMR. Even in adulthood, when muscles are well developed, the four metabolically most active organs account for two-thirds of the BMR.

Whatever its individual departures from large-scale means, *basal metabolism* follows a well-understood lifetime course: it starts with a brief spike in infancy, from about 2.3 watts per kilogram of body weight at birth to around 2.7 watts three to six months later, then it follows a steep decline to about half of the peak value by the eighteenth year before stabilizing for the next four decades. Resuming the decline at around the age of sixty, it falls to just around one watt per kilogram during the seventies. A final reminder: an individual's BMR should not be mistaken for the minimal survival requirement. Metabolic response to food, and energy needed to maintain basic personal hygiene will increase it by at least 15–20 percent.

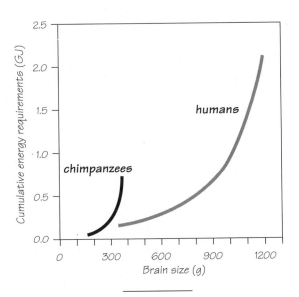

Huge differences in energy cost of encephalization
in chimpanzees and in humans.

Thermoregulation

Differences in *basal metabolism* — typical rates are somewhat higher in Northern Europe and among North American Indians and Inuit — have little to do with human adaptation to cold. Highest rates of *basal metabolism* were recorded among Arctic hunters, but that was due more to diets high in *lipids* and protein than to adaptive adjustments. Traditional multilayered Inuit clothes create an almost tropical microclimate, making it a greater challenge to dissipate heat during hard work rather than to conserve it.

Vasoconstriction is a useful adaptive response to cold nights in warmer climates. Naked Australian Aborigines could sleep without elevating their BMRs by reducing blood flow into skin and extremities, but this would be insufficient in colder climates. In humans, conservation of body heat has always been primarily a matter of intelligent

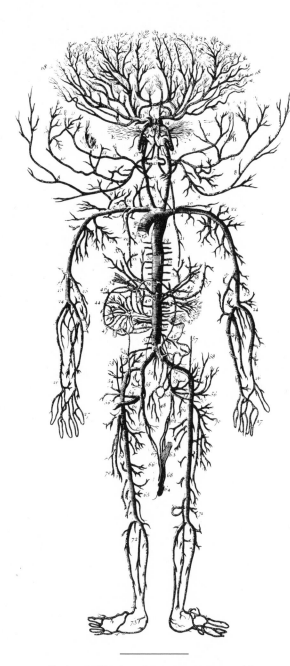

The human blood-circulation system is essential for effective thermoregulation: Vasoconstriction and dilation are its most important responses.

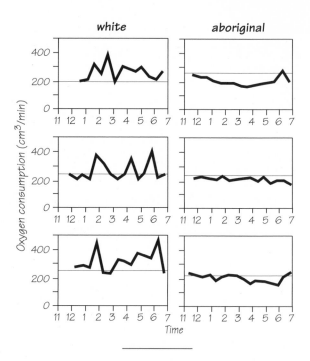

Vasoconstriction helped Australian Aborigines to have an uninterrupted sleep outdoors, while white men repeatedly woke up shivering.

extrasomatic solutions (clothes, shelter, fire) rather than one of biophysical adaptations.

Not so with human response to heat. Of course, protective microenvironments have been also important, ranging from counterintuitively dark-colored clothes (loosely worn, they absorb a large share of incoming radiation, which they then lose by convection without heating the body) to various passive cooling arrangements in traditional buildings. A highly effective extrasomatic adaptation to heat came only with the post-1950 diffusion of air conditioning—but outdoors we still have to rely on our innate thermoregulatory adjustments.

The initial response is to dilate peripheral blood vessels and to shift additional blood from internal to superficial veins. Then, usually when skin temperature

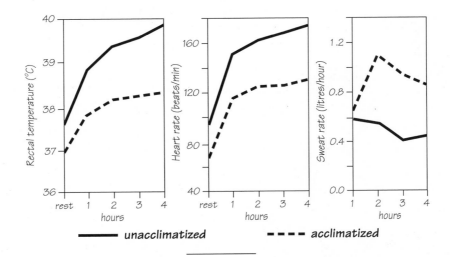

Substantial differences in temperature, heart rate and sweating
rate between acclimatized and unacclimatized Europeans
working in South African mines.

approaches 35°C, we begin to sweat, and during heavy work this response can be considerably more effective than in other mammalian species. A **horse** can perspire at a rate of 100 g/m² a hour, a camel can lose up to 250 g/m² — but a man can average more than 500 g/m² per hour. Without sweating average person would lose, about equally through respiration and skin diffusion, 12 W/m² of body surface, or just over 20 W in total. In contrast, hourly perspiration rate of 500 g/m² equates to heat loss of between 550–625 W for adults!

Best acclimatized individuals can perspire up to 1,100 g/m² per hour, an equivalent of 1,390 W. Such **thermoregulation** suffices to prevent a dangerous rise of core body temperature even in extremely hard-working individuals, but they require adequate rehydration. Temporary partial dehydration is common during heavy work or long footraces, and it causes no problems as long as the water deficit is made up within the next day.

High sweating capacities are genetically encoded and are retained even by populations that have lived for millennia in colder climates. Europeans, who initially respond to high heat by dangerously rising core temperatures and heartbeats approaching tolerable maxima, will start matching sweating rates of highly acclimatized tropical natives in only about ten days! Our ability to cope with heat by sweating must be seen — together with bipedalism, hairlessness, large brain, and symbolic linguistic ability — as one of the key defining human traits. Without it we could not have become either successful warm-climate **hunters and gatherers,** or, much later, builders of far-flung empires.

Pregnancy and Lactation

Considering its usual outcome — the miracle of a new baby — **pregnancy** is an astonishing energy bargain. Well-nourished women of rich countries add, on the average, about twelve kilograms of new tissues during the 280 days of normal **pregnancy.** The baby's birth weight (mean of 3.3 kg) and four to five kilograms of fat reserves for **lactation** account for most of the gain, with the rest in placenta

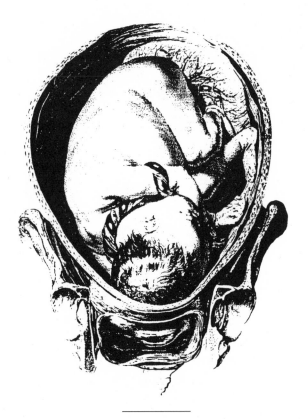

Full-term fetus *in utero* from Smellie's anatomical atlas (1754).

ment. ***Basal metabolism*** of active, healthy young women goes up by no more than 15–20 percent, an increase equivalent to food energy in just four slices of toasted whole-wheat ***bread*** a day!

Lactation costs vary considerably with the volume and duration of ***milk*** production. Average energy content of human ***milk*** is 3.2 kJ/g. Assuming a daily mean production of 800 mL for six months of ***lactation,*** the total energy cost of ***milk*** would be 520 MJ. But a typical Western woman would start her breastfeeding with some 150 MJ of fat reserves which will be converted to ***milk.*** Consequently, the actual postpartum energy cost of ***lactation*** would be only 370 MJ, an increment equal to about 25 percent of the normal food intake.

But these changes are not a universal norm. Impeccable studies in a number of Asian, African, and Latin American countries have shown that many women in poor rural

and in associated maternal ***growth.*** Further energy cost comes from the rise in ***basal metabolism*** and from increased cardiovascular and respiratory effort, especially during the last trimester. The total energy cost of such a ***pregnancy*** is about 335 MJ, or just 185 MJ when the cost of fat reserves is charged against the requirements of ***lactation.***

The larger total prorates to almost 27 kJ/g of weight gain, a higher rate than the cost of synthesizing new anthropomass during the years of childhood and adolescent ***growth.*** Still, the energy cost of a typical rich world's pregnancy represents only a modest additional food require-

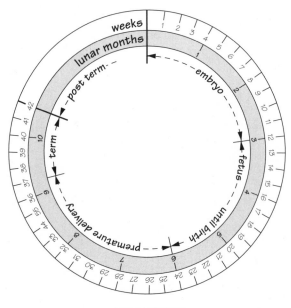

Stages of pregnancy.

Breastfeeding woman from Peter Brueghel
the Elder's *The Rich Kitchen* (1563).

foodstuffs are transformed with admirable efficiency (maxima go up to 97 percent!) into the nourishing fluid superior to any substitute. Breastfeeding not only lowers the frequency of several neonate diseases (including obesity, rachitis, and vitamin deficiencies), but it is also immunoprotective and confers important mental and neural advantages.

Human Growth

Normal infants grow fastest during the first nine months of life, when their weight gains are almost perfectly linear; those with smallest birth weights grow, on the average, up to 20 percent faster than the heaviest newborns. Afterward the lines of body weight plotted against age start curving gently for both girls and boys. By the end of their first year healthy infants almost triple their weight. Representative means differ both among nations and within countries over time. Standard North American charts show boys growing from 3.3 to 10.2 kilograms, and girls from 3.2 to 9.5 kilograms. Daily energy needs during this *growth* spurt average 430 kJ (or 5 W) per kilogram. Then the needs decline, first very slowly to around 400 kJ (4.6 W) by the age of four to five and afterward much faster to about 260 kJ per kilogram for girls and 300 kJ for boys.

Energy needs are always dominated by *basal metabolism: growth* needs peak at only about a third of all food intake during the first month of life, a storage rate of around 8 W. By the end of the first year the share of energy claimed by *growth* is sharply down to just between 5–6 percent, and by the end of the decade it is mere 2 percent of all food energy. *Growth* spurt during the early teens may briefly double that share, but by the age of sixteen or seventeen the rate is back to just around 2 percent, and falling fast. By the age of twenty the rate is down to a small fraction of one percent and it stays there for decades in order to repair and regrow adult tissues ranging

areas have extremely low energy costs of both *pregnancy and lactation.* Compared to Western expectations, their energy shortfalls are up to about 3 MJ/day even if they would just sleep and rest, and up to 4 MJ considering their heavy *labor.* These women, giving birth to healthy children, maintain genuine energy balance on what seem to be incredibly low levels of food intake, commonly 20–40 and even close to 50 percent below the expected requirement! And among the Kauls of New Guinea, British researchers found no difference in average food-energy intakes of nonpregnant and nonlactating and pregnant and lactating women!

Higher metabolic efficiencies of these women, adaptations to limited food intake, are the best explanation for these surprising realities. Equally notable is the relatively small effect of maternal nutrition on the quality of *milk.* Often even its quantity is not much lowered as *nutrients* in the plainest, and not infrequently barely adequate,

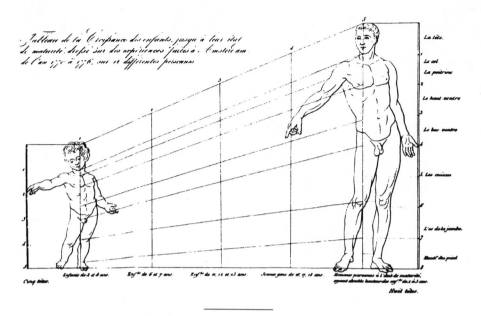

Proportions of human growth from the age of two years
to maturity, from a French publication of 1833.

from hair and nails, to constantly shed skin and intestinal lining.

Finding the efficiency of infant *growth* is not easy, because the inputs depend on the shares of stored protein and *lipids:* proteins need always more energy. Average rate of 21 kJ/g is the most commonly assumed energy cost of *growth* in young children. As the newborns average about 14 percent of fat and 20 percent muscle, the average energy content of their tissues would be about 10 kJ/g, and the overall efficiency of infant *growth* would be close to 50 percent. Other published *growth* rates range from lows of just below 15 kJ/g for infants recovering from malnutrition to more than 30 kJ/g for adults during overfeeding experiments.

Adult *growth* is greatly influenced by human sexual dimorphism exhibited in different rates of fat (adipose tissue) storage. After adolescence, the body's fat content increases steadily in both sexes, but the difference widens with age. *Lipids* make up about 15 percent of body weight in young Western adult males, but about 27 percent in females; by the seventh decade of life this disparity widens to 23 versus 36 percent. On the average, females are adding fat at rates of 0.3–0.4 kg/year, men at only 0.15–0.25 kg/year. This trend is accompanied by the loss of lean body mass.

Muscles are just over 50 percent of weight in young men, 40 percent in women. After the third decade the male's greater lean mass is lost more rapidly (2–3 kg per decade) than the female's musculature (about 1.5 kg per decade), and people over seventy years average about 40 percent less muscle than they had as young adults. This loss is an inexorable sign of physical aging even in those men and women who are in excellent health and who had avoided a significant fat increase.

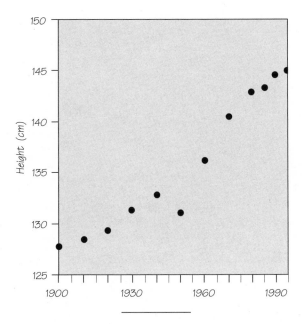

Better nutrition has resulted in steadily increasing heights of Japanese boys (average shown is for eleven years of age).

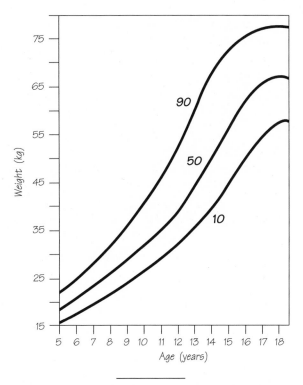

Growth curves, shown as tenth, fiftieth and ninetieth percentiles of body weights, of Canadian boys during the 1960s.

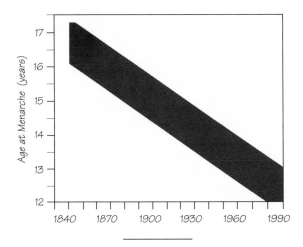

Better nutrition has also lowered the age of menarche throughout the affluent world from about seventeen years in 1850 to less than thirteen years in 1990.

Walking and Running

Walking is a key human trait—and it is also the most frequent activity requiring considerable energy inputs. More studies of energy expenditure are available for various modes of *walking* than for any other activity. Plots of their results show that at slow speeds the power needed for *walking* increases exponentially, but that the relationship at higher speeds is almost perfectly linear. Power inputs also vary with sex, age, body weight, load, slope, and surface quality.

Slim adult females have the lowest *walking* expenditures, but the rate is only about 10 percent below the male mean. Energy markups for lighter loads (up to about

Photographs of a running man from Eadweard Muybridge's
pioneering studies of human motion published in 1887.

15 kg) carried on a level are relatively small, similar to the cost of carrying a bit of additional body weight. Energy needs of *walking* uphill go up with both the gradient and the speed of the ascent, and numerous measurements indicate almost perfectly linear rise for a wide range of speeds and inclines. Uneven surfaces increase the input by up to one-third; loose sand or swampy ground can double it compared to unencumbered level *walking.* Total energy expenditures thus range from less than two hundred watts for slow level strolls to almost one kilowatts for strenuous *walking* uphill with a fairly heavy load, or two to ten times the rate of *basal metabolism.*

Energy cost of *walking* displays a distinctly U-shaped trend as it falls from maxima of close to 400 J/m for very slow walk to the minimum of about 230 J/m at the speed of 1.3 meters per second, and then it rises in an almost perfectly mirror-like fashion. Consequently, human terrestrial locomotion is most efficient when *walking* at between 4.5 and 5 kilometers per hour—but most people

walk either much slower or considerably faster. People in small villages often walk with speeds below three kilometers per hour, while many hurrying pedestrians in large metropolitan areas commonly exceed six kilometers per hour.

Although record race-*walking* speeds surpass four meters per second, these feats are achieved by peculiar rotations of pelvis about its vertical and horizontal axes. Normal *walking* is transformed into *running* once the speed reaches about 2–2.5 meters per second, or 8.3 kilometers per hour. Above that speed *running* has lower energy cost than *walking,* and while the cost of *walking* keeps on increasing with speeds above 1.3 meters per second, the cost of *running* does not vary significantly with speeds between 2.3 and 6 meters per second. This uncoupling of energy cost from speed is a unique human ability that can be explained by a combination of bipedalism and efficient heat dissipation.

Quadrupeds have optimum speeds for different gaits (for example, the horse's walk, trot, and canter) because

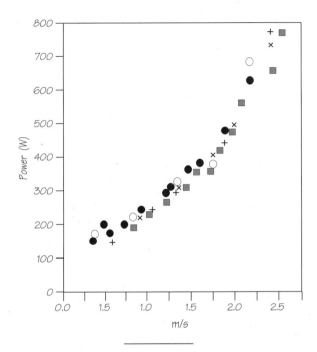

Energy expenditures of walking on the level increase linearly for speeds up to about 7 km/h (almost 2 m/s).

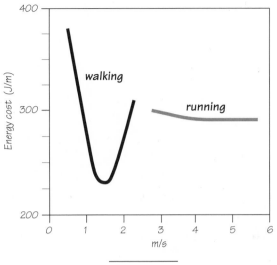

Energy cost of walking has a clear U-shaped relationship with speed, while energy requirements for running remain basically constant for speeds between 3 and 5.5 m/s.

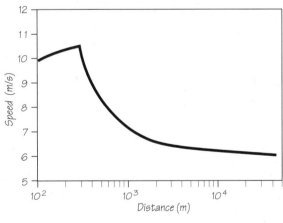

Maximum running speed as a function of distance. Up to about three hundred meters, runners can rely on stored energy converted anaerobically; above that level their speeds must decline to conform to the slower aerobic balance.

their ventilation is limited to one breath per locomotor cycle: their thorax bones and muscles must absorb the impact on the front limbs as the dorso-ventral binding rhythmically compresses and expands the thorax space. But human breathing can vary relative to stride frequency, giving us an option to run at a variety of speeds with basically unchanged energy costs. As for the *thermoregulation,* our ability to dispose of metabolic heat surpasses that of any other mammal. Combination of variable-speed *running* and excellent heat dissipation makes it possible for people to outperform some of the fastest mammals. This ability was exploited by many hunting cultures: North American Indians used to run down deer and pronghorn antelopes; Kalahari Basarwa chased duikers, gemsbok, and zebras; and some Australian Aborigines pursued kangaroos.

These endurance runs have been repeatedly surpassed in classical marathons (42.2 km) and in modern ultrarunning races (100–160 km). Since the 1970s the best athletes have been *running* marathons at a faster pace—at speeds of nearly 5.5 meters per second—than the record ten-kilometer run of the early 1940s. Ultrarunning races are now covered at speeds of three to four meters per second. All long-distance speeds are limited by two organismic rates: by the uptake of oxygen in the lungs and by the distribution of the blood by the heart. In contrast, sprinters derive most of their energy from anaerobic sources. Record sprint speeds translate to a mean of 10.2 meters per second, a feat achievable thanks to a rapid use of creatine phosphate and anaerobic glycolysis. Recent speed improvements in sprint races have been, obviously, very small in absolute terms, but very similar to those of endurance races in relative terms. And although record speeds of female runners are still considerably behind the best male achievements, women's rates of improvement have been appreciably faster.

Typical power inputs in *running* range between 700 and 1,400 watts, corresponding to metabolic scope of between 10 and 20. Training can raise these rates, mainly by

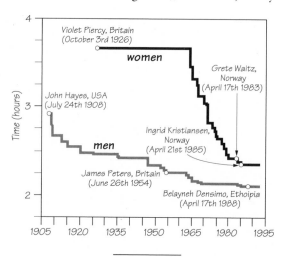

History of world records for the marathon.

increasing blood volume (roughly one mL for each mL of a aerobic capacity) and hence the cardiac output. Top rates for elite athletes can go over two kilowatts, that is, about twenty-five times the BMR. This metabolic scope is higher than for rabbits but lower than for wolves and dogs, the best mammalian runners.

Labor and Leisure

Since the 1890s physiologists have accumulated thousands of values for energy costs of specific tasks in scores of occupations as well as in just about every recreation activity. Most of them have been determined by simple respirometry: The energy equivalent of one liter of consumed oxygen averages about 20 kJ per gram of oxidized *nutrients*. A simple way to express these costs is to compare them to *basal metabolism*. The stimulating effect of the last meal and some tossing will push sleep's energy cost 5–15 percent above the basal metabolic rate (BMR). Sitting and standing require the deployment of many muscles to maintain various postures: typical markups for sitting are 15–20 percent, while standing requires 1.3–1.5 times the BMR. The huge assortment of occupational tasks is best reviewed by typical levels of exertion. This approach brings many surprising results.

Thinking is an enormous energy bargain. The adult brain claims about a fifth of the BMR, but even the hardest brainstorming makes little difference to that fixed rate: it requires no more than about four watts, equal to around 5 percent of a typical BMR. All modern science is thus a matter of light energy expenditures—excepting, naturally, climbs into rain forest canopies or crawling through constricted caves.

Expectedly, nearly all tasks in the service sector are light exertions. This huge category embraces not only all deskbound jobs, but also virtually all cleaning, retail, repair, teaching, hospital, or restaurant tasks, as well as most transport services: driving a truck may actually require

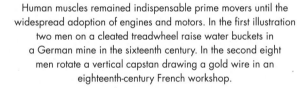

Human muscles remained indispensable prime movers until the widespread adoption of engines and motors. In the first illustration two men on a cleated treadwheel raise water buckets in a German mine in the sixteenth century. In the second eight men rotate a vertical capstan drawing a gold wire in an eighteenth-century French workshop.

less energy than typing sixty words a minute! Modern manufacturing—be it assembling computers or making trucks—is also overwhelmingly a matter of light work, as is nearly all mechanized farming. Machines have also shifted all but a few modern construction tasks into the light exertion class.

In contrast, traditional farming still involves a great deal of moderate and heavy *labor*. The most common heavy *labor* tasks are plowing, weeding, hoeing, transplanting, digging, canal cleaning, brush clearing, and mowing. Similarly, traditional net fishing, tree felling, and mining entail heavy—and often very heavy—energy expenditure. Naturally, total daily expenditures depend on

the duration of individual tasks. Moderately taxing, steady work may be thus more demanding than jobs requiring occasional heavy exertion. Underground coal miners do many tasks calling for power of 550–600 watts, but they may spend half of their day getting to and back from the coal face and in periodical resting, averaging no more than three hundred watts, a decidedly light activity. In traditional farming such disparities have also a clear seasonal flux: periods of intense heavy *labor* (planting, harvesting, plowing, digging) are followed by weeks of moderate work and extended rest.

Daily rates of energy expenditures range from as little as 6 MJ for older housewives to over 30 MJ for lumberjacks. Many ergometric studies put the best muscle efficiencies during extended, aerobic work at around 20 percent. Consequently, even a hard-working lumberjack accomplishes daily useful work equivalent to just 6 MJ. That much useful kinetic energy can be delivered, even when the overall conversion efficiency of an inanimate prime mover would be just 20 percent, by burning about seven hundred grams of *crude oil* or one kilogram of good *coal*. Clearly, human effort, even at its best, is a rather unimpressive source of mechanical energy!

The energy cost of *leisure* activities is highly dependent on their intensity: Normally moderate exertions, ranging from badminton and canoeing to swimming and tennis, may be readily pushed into the heavy category. Individual high-energy sports include many track and field events, mountain climbing, cross-country *running,* rowing, and squash. Cycling is the fastest mode of human locomotion, and it has energized all human-powered record-setting machines on land, water, and in the air. Group sports with the highest energy costs are basketball and soccer.

Short (30–180 seconds) exertions are energized mostly by anaerobic glycolysis; aerobic recharge is the

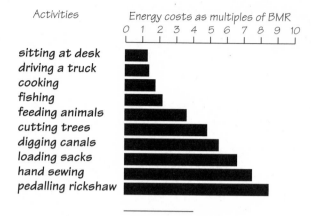

Gross energy expenditures of some activities expressed as multiples of basal metabolic rates for males.

main, though not the sole, energizer of sustained exertions (anaerobic breakdown of glycogen also contributes). Because the body's oxygen stores can support a heavy effort for less than half a minute, subsequent energy needs call for linear increases in oxygen uptake. The peak aerobic power reaches six hundred to nine hundred watts in mildly active adults, but it can top two kilowatts in elite athletes; it is slightly lower in females than in males of the same age, and in untrained individuals it declines steadily after the adolescence.

Healthy adults can easily work or exercise for several hours at 40–50 percent of their peak aerobic capacity, that is, at rates of three hundred to five hundred watts. In terms of total work accomplished, peak aerobic capacities are equivalent to 1.5–3.5 MJ for healthy adults, surpass 10 MJ for good athletes and reach an astonishing 45 MJ for best long-distance runners and skiers!

Nutrients

Human life depends on the digestion of about fifty essential *nutrients* ranging from complex organic compounds to mineral elements. *Nutrients* consumed daily in larg-

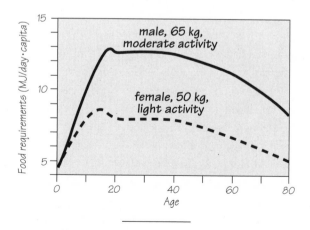

Food requirements vary with age, sex, and level of activity.

est quantities—carbohydrates, proteins and *lipids*—are needed to supply energy for *basal metabolism, growth,* and *labor and leisure* activities, but even these compounds have qualitative roles that enhance, or even greatly surpass, their energy contributions. Vitamins and minerals have no energy value and, compared to the three macronutrients, are needed only in minuscule amounts (ranging from a few grams a day for alkaline elements to a few milligrams for some vitamins), yet their adequate intake is essential for healthy life. Although it makes no sense to rank *nutrients* as to their overall importance, there is no doubt which ones make the greatest contribution to human energy requirements.

Carbohydrates not only ranked first, but they also accounted for the bulk of digested food in every traditional society; they still provide more than three-quarters of all food energy throughout the poor world, but in the richest countries their share has fallen below 50 percent. People eat carbohydrates in a wide variety of processed cereal and leguminous *grains*—most commonly as *bread,* pasta, steamed rice, or various gruels, stews, and fermented products—in tubers, fruits, and vegetables, or as sugar, honey, or concentrated tree sap.

No matter if they come as complex starches (polysaccharides made up of thousands of glucose molecules) or simpler sugars (monosaccharides fructose and glucose, and disaccharide sucrose), carbohydrate energy content is

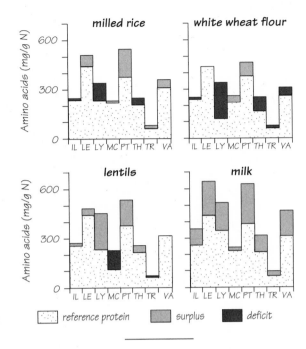

Amino acid composition of rice, wheat flour, lentils, and milk.

nutrients	energy content (kJ/g)	
	total	actually available
carbohydrates	17.4	17.0
lipids	39.0	38.0
proteins	23.0	17.2
ethanol	29.3	29.3

Energy content of nutrients.

just 17 kJ/g. Unfortunately, most of the biosphere's carbohydrates cannot be digested by humans: we lack the enzymes needed to break down wood's lignin, cellulose, and hemicellulose. But since the 1960s we have come to realize the nutritional importance of these indigestible carbohydrates. These dietary fibres—richly present in whole seeds, whole flours, fruits, and vegetables—are needed daily only in small amounts, but they are a critical part of proper nutrition: We consumed them in much higher quantities during millions of years of hominid evolution, and we still need them to prevent, or to reduce, the incidence of many diseases.

Proteins, with 23 kJ/g, are about a third more energy-dense than carbohydrates—but their principal role in human nutrition is not as sources of energy but rather as suppliers of essential amino acids needed to build body

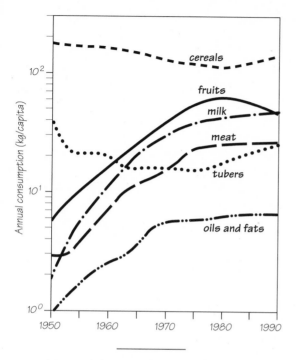

Nutritional changes in post–World War II Japan.

activities, to release hormones, and to build cell membranes. Concentrated *lipids* are eaten as various plant oils, butter, and lard. Lipid content of *meat,* fish, and dairy products ranges from trace amounts in skinless poultry breasts and skimmed *milk* to much more than 50 percent in fatty pork cuts or *double-crème* cheeses.

No macronutrients are completely digestible. Healthy people on balanced diets can derive up to 99 percent of available energy from carbohydrates, about 95 percent from fats, and no more than 92 percent from proteins; in addition, more than one-fifth of consumed protein (about 5.2 kJ/g) is voided in urine. Net energies actually available for *basal metabolism, growth,* and *labor and leisure* activities are thus basically identical to the gross energy density of carbohydrates, are only a bit lower for *lipids,* but amount to less than three-quarters of the gross value for proteins. This adjustment matters because standard food composition lists convert energy content of foodstuffs by using net metabolic factors, not gross energy contents.

Micronutrients—minerals and vitamins—provide no food energy, but their importance in building healthy tissues and maintaining the complex biochemistry of human metabolism is irreplaceable. Consequently, diets with a surfeit of energy may still lead to serious cases of malnutrition. Intakes of more than forty micronutrients are required for healthy living. Minerals needed in largest amounts are calcium, phosphorus, and iron; vitamin requirements are highest for C, B complex, and D.

Grains

Several energy advantages explain the dominance of cereal and leguminous *grains* in human diets. Above all, *grains* combine fairly high yields with relatively high energy density. Cereal yields have been always appreciably higher (commonly at least two times) than the harvests of leguminous seeds, but legumes, associated with nitrogen-

tissues. Normal human *growth* is impossible without digesting eleven essential amino acids which are then reorganized into protein in body structures (in muscles, bones, and internal organs) as well as in a variety of irreplaceable biochemical carriers, catalysts and protectors (antibodies, enzymes, hormones). Proteins come complete—that is, including all essential amino acids in requisite proportions—in all animal foods as well as in mushrooms, and incomplete (with one or more amino acids deficient) in plant foods (mainly in leguminous and cereal *grains* and in nuts).

Lipids—fats and oils—are by far the most energy-dense *nutrients* with 39 kJ/g. They contain three essential fatty acids whose digestion makes it possible to carry fat-soluble vitamins (A, D, E, and K) and to synthesize prostaglandins needed to regulate gastric and smooth-muscle

The three great staples of European cereal cultivation: wheat, rye, and barley.

Moreover, the complementarity of cereals (deficient in lysine) and legumes (low in methionine) makes it possible to consume all essential amino acids in *grains*. Traditional, overwhelmingly vegetarian societies practiced this complementarity—millennia before we understood its biochemical basis—by combining staple cereals with soybeans and beans in East Asia; with lentils, beans, and peas in India, the Middle East, and Europe; with peanuts and cowpeas in West Africa; and with beans in the Americas. Oils are present in most cultivated varieties in modest amounts, but their content is high in corn, peanuts, and soybeans.

Other advantages, nutritional and logistic, include relatively easy harvesting and simple basic processing of grain crops; low moisture content, which makes mature cereals and legumes suitable for long-term storage; and the culinary versatility of *grains*. The ease of harvesting and processing—especially when compared to tubers, roots, and nuts, which supplied large shares of food energy for all gatherers—was an important contributing factor in the emergence of *grains* as staples of virtually all settled societies.

fixing bacteria, are easier to grow and require hardly any fertilizer *nitrogen.* Energy density of mature *grains* differs by less than 10 percent among the main cultivated genera: with moisture content at around 14 percent (the maximum necessary for safe storage) they contain about 15 MJ/kg. This is roughly five times higher than for tubers, and equal to moderately fatty cuts of *meat.*

All principal *nutrients* contribute to this rather high energy content. Carbohydrates are always dominant, present mostly as highly digestible polysaccharides (starches). Protein content of cereals ranges widely, from just 7 percent for some rices to 16 percent for the Andean *quinoa*. A typical mean of around 10 percent is about five times higher than that of tubers, and more than an order of magnitude above the values for vegetables. Leguminous *grains* average over 20 percent of protein and soybeans over 40 percent.

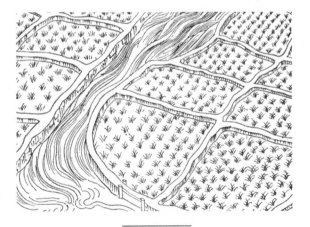

Cultivation of rice, Asia's dominant staple, is shown here in a classical Chinese illustration.

milled and reconstituted as solids (breakfast cereals), or as fermented products (soybean-based bean curd, bean paste, and soy sauce are the best known possibilities). Milling, the dominant mode of cereal processing, has little effect on their overall energy density; it increases palatability, but it also decreases nutritional value by removing vitamins and minerals present in the seed's outer layer, and it reduces the beneficial intake of indigestible fibre.

As for the cooking methods, *grains* can be boiled in rough or elaborate gruels (from Scottish oatmeal porridge and Russian buckwheat *kasha* to satiny East Asian rice *congees* with floating condiments); cooked in hot water as

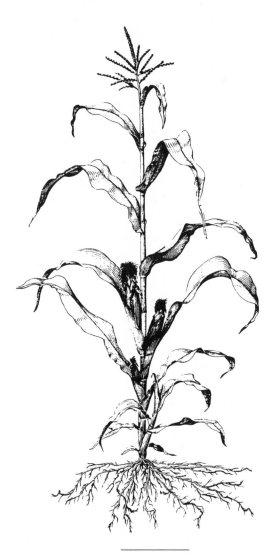

Corn, the New World's most important addition to global grain harvests.

The versatility of *grains* as food is obvious from listing main processing and preparation options. *Grains* can be eaten whole (all legumes), crushed (oatmeal), whole after removing the outside layer by milling (rice), milled to various degrees of fineness in flours (wheat above all),

Winnowing—separation of chaff from grain—from a seventeenth-century Chinese encyclopedia.

with the almost endless varieties of European and Asian pasta; steamed; and baked as *bread,* for millennia the leading staple of Western civilizations, and as a huge assortment of salty and an even wider variety of sweet pastries crowned by French *gateaux* and Viennese *Torten.*

The regional or local dominance of particular *grains* is largely a matter of environmental possibilities, agronomic traditions, and taste preferences. Wheat spread from the Near East to become the most widely cultivated cereal; rice, originally from the Southeast Asia, is now the world's largest grain crop; corn was introduced from North America to every continent, and it is now a leading feed grain; soybeans are as well, having spread from China.

Cultivation of many coarse *grains*—millets, barley, rye, buckwheat—has been in a global decline. So has been, except for soybeans and Indian lentils, the planting of leguminous *grains.* Although high in protein, they have often low palatability and are difficult to digest. As soon as societies get richer one of the most notable nutritional shifts is their declining consumption of legumes.

Per capita consumption of cereal staples also declines with modernization. In traditional societies they commonly supplied nine-tenths of all food energy. Now they provide about three-quarters in China, less than half in North America. But the total cultivation of *grains* in rich societies has grown: instead of eating them directly, rich countries now channel most of their cereals through animals to produce more animal foods. Globally nearly two-fifths of harvested *grains* were fed to animals in the early 1990s, and the share is above 60 percent in rich countries.

Bread

Several cereals, and many nongrain ingredients, have been used to make *bread.* American *tortillas* contain ground corn with lime water. In northern parts of Europe heavy, dark *bread* was often made only from rye flour with the

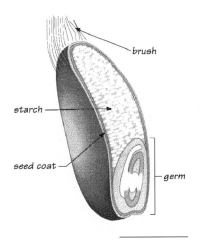

Wheat grain.

addition of such coarse *grains* as oats, barley, and buckwheat. Lentils or cauliflower may be mixed into Indian *chapatis.* Bananas and molasses, *milk* and eggs, nuts and dried fruits can be used to enrich fancy *breads,* spices and herbs to flavor even the plain ones. But wheat flour remains the quintessential ingredient for the world's most important baked staple.

Two of its major proteins, glutenin and gliadin, are unique— not nutritionally, but because of their physical properties. When combined with water they form a gluten complex that is sufficiently elastic to permit stretching and shaping of an unleavened and rising of a leavened dough, and yet strong enough to retain carbon dioxide bubbles formed during the yeast fermentation. Leavened *bread*— a fermented, kneaded, and baked mixture of flour, water, salt, and *Saccharomyces cerevisiae* yeast—was such a dominant staple in traditional European societies that medieval Latin writings commonly referred to all other foodstuffs collectively as *companagium,* accompaniments to the ubiquitous *panis* served with all meals.

Laborious grinding of flour was one of the principal reasons for introducing first inanimate prime movers,

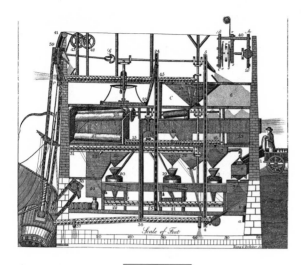

Breadmaking became cheaper with the introduction of first automated mills. This is Oliver Evans's pioneering design from 1785.

commonly home-made, baked stuck to the sides of clay ovens, grilled (Mexican *tortillas*), fried in oil (Indian *puri*), or steamed (Chinese *mantou*).

The energy cost of **bread** depends on the scale of production. At the retail level, American white loaves from a large commercial bakery using the standard sponge-and-dough process will need around 7 MJ/kg, with the cost split about 2:1 between production and delivery to stores. Baking itself will need less than 20 percent of that total. European costs are very similar, but shorter distribution distances make for lower transportation costs. The specific energy cost of home baking can be up to five times higher — but there are no indirect energy costs for packaging and distribution. And, not surprisingly, the process displays striking economies of scale: three **breads** in oven

waterwheels and *windmills.* Flour mill operations (grinding, separation, sifting, and bagging) were fully mechanized only after 1800; by 1900 *steel* rollers had almost totally displaced elaborately dressed millstones. Yeast, traditionally gathered from beer or wine vats, is now mass-produced from cultures grown on molasses.

Most standard breadmaking recipes call, in mass terms, for a ratio of 1:1.7 for water and flour. Because the flour itself has about 12 percent of moisture, fresh doughs are roughly 45 percent water. Most of the *wood, coal,* or *electricity* used in baking goes into evaporating part of that water and reducing **bread**'s moisture content to as low as 30 percent in light baguettes, and up to 36 percent in heavier whole-wheat and rye loaves. Fresh **bread** has relatively high energy density, equal to about 75 percent of the identical mass of flour. Leavened **bread** is now mostly mass-produced in commercial ovens, but Asian, African, and Latin American unleavened **breads** are still

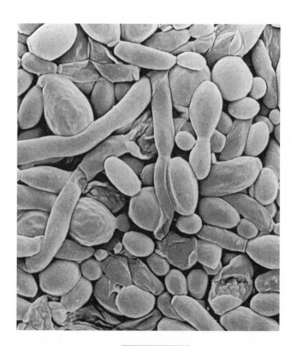

Saccharomyces yeast.

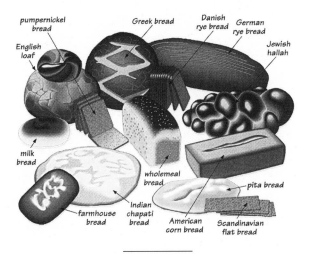

pumpernickel bread
Greek bread
Danish rye bread
German rye bread
Jewish hallah
English loaf
milk bread
wholemeal bread
pita bread
farmhouse bread
Indian chapati bread
American corn bread
Scandinavian flat bread

Variety of breads.

need about 50 percent less energy per kilogram than a single loaf.

Lipids and Meat

Animal fats and plant oils have the highest energy density of all *nutrients.* This attribute made them highly regarded in every traditional society. Hunting societies preferred killing the largest mammals not just because those animals provided plenty of *meat,* but also because that *meat,* in contrast to that of smaller creatures, was also uncommonly rich in fat. Bison provided twice as much energy per unit weight as an elk or a deer, and the same difference applied to elephants compared to even the largest antelopes.

Similarly, the high fat content of Pacific baleen whales and salmon provided the energetic foundation for settled and relatively complex societies of the Pacific Northwest: precontact settlements totaled up to several thousand people, concentrations unattainable by hunting lean *meat.* The most commonly landed immature whales averaged nearly twelve tonnes, with blubber rated at about 36 MJ/kg, and *mukluk* (skin and blubber) at 22 MJ/kg, while

mass migrations of salmon yielded catch with about 15 percent of body mass in fat, or nearly three times higher energy density than in cod.

In overwhelmingly vegetarian agricultural societies, diets rich in *lipids* not only gave the satisfactory feeling of satiety but also reflected the consumer's high social status. For most peasants in preindustrial societies, oils, butter, lard, and fatty *meat* were only occasional luxuries, supplying commonly less than 10 percent of all food energy, a pattern that also characterized typical diets during the early stages of industrialization.

Modern societies have clearly overcompensated for this deficiency: *lipids* now make between a third and four-fifths of total food-energy intake in the United States, Canada, and most European nations, a clear nutritional excess. The best dietary guidelines recommend that no more than 30 percent of food energy should come from *lipids.* In recent years there has been some reduction of fat intakes throughout the affluent world, especially because of

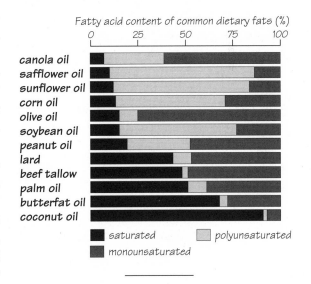

Fatty acid content of common dietary fats (%)

canola oil
safflower oil
sunflower oil
corn oil
olive oil
soybean oil
peanut oil
lard
beef tallow
palm oil
butterfat oil
coconut oil

saturated   polyunsaturated
monounsaturated

Fatty acids in common dietary lipids.

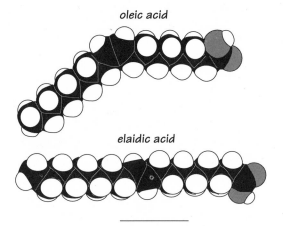

oleic acid

elaidic acid

Examples of unsaturated (curved) and
saturated (straight) fatty acids.

the warnings about health perils of high fat diets in general, and their effects on higher incidence of premature coronary heart disease in particular.

These warnings have turned *lipids* into a questionable treat for many people — but *lipids* provide much more than the satisfying feeling of a filling meal. Two of their constituent fatty acids — linoleic and linolenic — are essential macronutrients needed not for their high energy content but for their critical biostructural functions. These acids cannot be synthesized by vertebrates, but all cell membranes are made largely of *lipids,* as are the myelin sheaths around nerve fibers. Consequently, adequate intake of essential fatty acids is especially important during infancy: about 60 percent of the brain mass increase after birth is structural fat. Fatty acids are also precursors of prostaglandins, which regulate gastric functions, release of hormones, and the activity of smooth muscles, and they carry vitamins A, D, E, and K, which are not soluble in water.

Both linoleic and linolenic acids are polyunsaturated compounds with, respectively, two and three double

bonds that make their long carbon chains (both have 18 C) curved and keep the acids liquid. Most plant oils, ranging from corn to peanuts and from sunflower to soybeans, are high in polyunsaturated acids. In contrast, saturated fatty acids — myristic, palmitic, and stearic, with, respectively, 14, 16, and 18 carbons — have no double bonds, and their straight chains form solid structures of animal fats, mostly consumed as rendered lard and tallow, and they are also abundant in coconut and palm oil. Monounsaturated oleic acid has a single double bond, enough to keep olive oil liquid.

Hydrogenation has been used since the late nineteenth century to straighten carbon chains — to have their *cis* double bonds changed to *trans* double bonds — and to make mostly unsaturated oil into a solid fat. After 1950 this transformation seemed to be the best solution to reconcile widespread human preference for fatty diets with the need to limit intake of solid saturated animal fats associated with higher incidence of coronary heart disease: butter consumption in rich countries fell sharply as margarine sales soared.

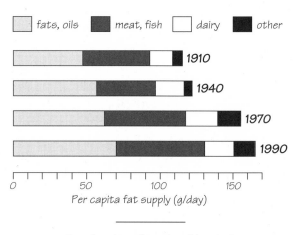

Growth and transformation of America's
dietary lipid intakes, 1909–1990.

Trans fatty acids now supply as much as 15 percent of all energy from fat (or more than 5 percent of all food energy) for some people in Western countries, but growing evidence has showed that at that level of intake they may have a number of adverse effects. Among the most worrisome ones is the lowering of high-density lipoproteins ("good" cholesterol), reduction of fat content in human *milk*, increased risk for diabetes, and alterations in the activities of the enzyme system metabolizing chemical carcinogens and medications. Consequently, the best dietary advice seems to be to reduce intake of all *lipids* and to eat moderate quantities of both saturated animal fats, and mono- and polyunsaturated plant fats—as long as they are not hydrogenated.

Milk

For an emulsion containing more than 85 percent of water, *milk* has an unusually high energy density. Human *milk*, whose energy content is almost 20 percent higher than that of cow's or goat's *milk*, has not only more digestible food energy than all common temperate climate fruits, but it also just edges white potatoes. Much like *grains*, every *milk* derives this high energy density from a nutritionally balanced combination of principal *nutrients*. With the exception of reindeer *milk*, which is exceedingly rich in both protein and *lipids*, all commonly drunk *milks* have less than 5 percent fat. Human *milk* has only about a third of cow's *milk* protein (just over 1 percent), but it is nearly twice as high in carbohydrates.

Breastfeeding gives undoubtedly the best nutritional start to a new life: *lactation* produces a virtually perfect food for infants. But animal *milk* is not necessarily an ideal food for older children and adults. Its sugar is lactose, a disaccharide composed of glucose and galactose. Digesting lactose is easy for all normal infants: they have enough lactase to hydrolyze it. But after weaning, hun-

milk	human	cow
energy (kJ/g)	3.2	2.7
water (%)	85	87
protein (%)	1.1	3.5
fat (%)	4.0	3.5
Ca (mg/100 g)	33	118
vitamin A (IU)	240	140

Comparison of human and cow milk.

dreds of millions of children around the world lose most or all of the enzyme, and the lactase deficiency prevents them from consuming lactose-rich diets without side effects ranging from unpleasant (intestinal cramps, flatulence) to painful (diarrhea, vomiting, nausea).

Lactose malabsorption has a complex spatial pattern. Its highest prevalence is among the people of traditional hunter-gatherer societies (Inuit, all American Indians, South African Basarwa) and in nonmilking farming cultures of sub-Saharan Africa and Southeast and East Asia. In contrast, people in Northwest Europe, North Africa, and the Near East have very high lactose tolerance. Disparities in the incidence of lactase deficiency are huge: well over 90 percent among Japanese and Chinese, much below 10 percent among Germans and Swedes.

Yet lactase deficiency does not prevent consumption of dairy products. They can be taken either in small quantities, or in larger amounts when fermented. *Lactobacillus* fermentation reduces lactose in yogurt by at least 30 percent. Unripe cheeses (ricotta, cottage cheese) lose commonly more than 70 percent, and the ripe ones—soft (Camembert or Roquefort) or hard (Edam or Cheddar)— contain merely a trace of the sugar.

With declining breastfeeding caused by urbanization and rising female employment, *milk* and dairy products

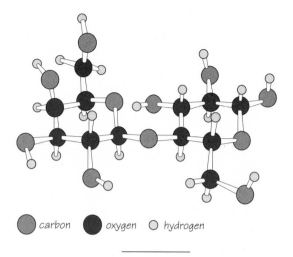

carbon · oxygen · hydrogen

Milk sugar, lactose, is composed of glucose and galactose.

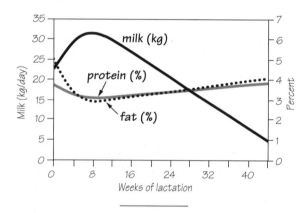

Changing Holstein cow milk production with advancing lactation.

have made surprisingly strong inroads even in societies where *milk* drinking was traditionally absent. Most notably, Japan's per capita consumption rose from nothing in 1945 to more than forty liters by 1990, and more *milk* is now drunk every year in China's large cities. Cow's *milk* accounts for nearly all of this expansion. Even avid drinkers of that smooth, sweetish emulsion do not find other

kinds of *milk* to be such gustatory delights. Acquired tastes include not only goat and sheep *milk* (in much of the Old World and in Latin America), but also water buffalo (mainly in monsoonal Asia) and camel (Arab world) *milk*. Mare and yak *milk* drinking is limited to parts of Central Asia, and reindeer *milk* is drunk in the northernmost reaches of Eurasia (from Scandinavian Lapland to Siberia).

Ethanol

Fermentation of alcohol from a variety of carbohydrates (most commonly from cereal *grains* and sweet fruits) has been done since antiquity — and the intoxicating effect, rather than any nutritional gain, is its main goal. Nevertheless, *ethanol*'s food-energy density compares favorably with values for the three main food *nutrients.* Its 29.3 kJ/g are some 70 percent above the density of digestible carbohydrates and proteins, and are equal to nearly 80 percent of the energy content of *lipids.* Curiously, usual servings of common alcoholic beverages accompanying food have nearly identical ethanol content regardless of their variety.

One glass of good beer (350 mL, 4 percent ethanol), table wine (120 mL, 12 percent ethanol) or a fortified aperitif or dessert wine (75 mL of sherry, Marsala, Cinzano, or Dubonnet containing 18 percent ethanol) — all have about 14 mL (or 11.5 grams) of alcohol. But their energy contents will not be identical. Because of different

Alcohol metabolism in human body.

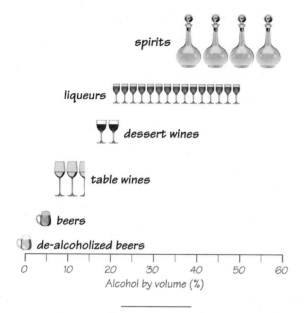

Alcohol content of common beverages.

other liquids, ingesting particular foods, or engaging in a strenuous exercise.

This rather sluggish constancy of *ethanol*'s metabolic breakdown means that while alcoholic beverages may supply a large share of everyday energy needs for habitual consumers, they can never cover the total daily requirement. An active, fifty-year old, healthy and not overweight (70 kilograms) French farmer will have to metabolize about 10.5 MJ of food energy to get through the day. His daily bottle of *vin ordinaire* will give him exactly one-fifth of that demand, his liver will easily process the *ethanol* load amounting to just over two-fifths of its maximum clearing rate — and the wine, unlike other alcohols, may give him significant cardiovascular protection so that his chances of heart attack will be lower than those for his distillate-drinking counterpart in Scotland.

carbohydrate shares — less than 4 percent in beer, up to 8 percent in cream sherries — a glass of beer will have about 540 kJ, a glass of table wine 350 kJ, and a glass of sweet sherry 470 kJ.

Complex or simpler sugars in these alcoholic beverages will be metabolized as any other carbohydrate — but ethanol can be transformed to acetaldehyde, the first step of its breakdown to acetic acid and one of the compounds responsible for hangover, only by alcohol dehydrogenase. This enzyme is present in the liver, and it is needed to handle small volumes of *ethanol* produced by normal digestion of carbohydrates and larger volumes generated by intestinal bacteria. The rate of this reaction is just around 0.1 gram of ethanol per kilogram of body weight per hour, that is between 150–260 kJ/hour (42–72 W) for most people. This rate cannot be speeded up either by drinking

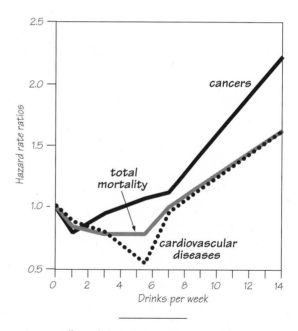

Effects of alcohol consumption on mortality.

In contrast, a terminal alcoholic of the same weight and age leading a collapsing life of minimum activity can get by with as little as 7.5 MJ of food energy per day. His maximum metabolic clearance of about 170 grams of *ethanol* per day corresponds to two-thirds of a bottle of 80-proof whiskey and supplies 5 MJ, still one-third short of his minimum energy needs. He may, of course, finish the whole bottle of the liquor every day, but that would only speed up his physical and mental deterioration without supplying more energy.

4

PREINDUSTRIAL SOCIETIES

Our ancestors spent more than nine-tenths of their existence as *hunters and gatherers* in activities that required many physical and mental adaptations and were also indispensable for the emergence of social complexity. But only in some coastal communities, tapping rich seasonal migrations of ocean fish and mammals, could foraging support high population densities and lead to a sedentary existence. On *grasslands* and in *forests* population densities of roaming foragers were hardly higher than those of their primate ancestors.

As their numbers rose, most foragers had to turn to an increasingly sedentary way of cropping. These agricultural practices began with *shifting cultivation.* In this least energy-intensive mode of crop production the cultivation of several crops of tubers, *grains,* or fruits on a patch of land cleared of natural vegetation (usually by fire) alternated with often long periods of fallow.

A further rise in population densities brought a variety of *traditional agricultures.* Throughout the Old World human groups shared domestication of animals, reliance on plowing, and the cultivation of staple cereal and leguminous *grains.* None of the New World's complex societies — Incas, Mayas, or Aztecs — acquired the first two traits, but their cropping was often more productive than harvests in Europe and Asia.

Many *cattle* breeds provided draft power, both for field work and for transportation. They also supplied *milk,* but they were too valuable to be slaughtered for *meat* unless they got very old. *Horses* were used extensively in war from antiquity — but they became superior draft animals only with the adoption of an efficient harness and iron horseshoes, and their widespread use in *traditional agriculture* came only when changes in cropping patterns provided enough concentrate feed.

Virtually all fuel in preindustrial societies came from *wood, charcoal, and straw.* For household cooking and heating, these fuels were burned inefficiently in a variety of fireplaces; enclosed stoves with chimneys are a surprisingly late innovation. Because of its high energy density, *charcoal* was the preferred fuel for smelting and processing metals, mainly *copper, iron, and steel,* and for firing bricks.

Most preindustrial *labor* was done by muscular exertions of people and animals. More powerful sources of kinetic energy—*waterwheels* and *windmills*—were invented only after millennia of settled societies, and in most such societies they made only marginal contributions. The capacities of these mechanical prime movers, used for many food processing and manufacturing tasks and also in raising water, grew only slowly.

Improvements in the typical performance of *sailships* were also very slow. Fundamental breakthrough came only at the beginning of the early modern era. At that time the combination of more maneuverable vessels with more accurate guns (made possible by advances in the smelting of *copper* and *iron* and by the invention of *gunpowder*) produced an energy converter of unprecedented speed, range, and destructive power that helped to usher a new era of world history.

Hunters and Gatherers

Only an uninformed view would not perceive tens of thousands of years of hominid foraging as a prolonged prelude to a truly sapient existence in increasingly complex civilizations. To continue the musical analogy, it was very much like acquiring a large ensemble of specialized instruments, fine-tuning them, and getting them ready to play ever more intricate scores. Development of all of the key characteristics distinguishing humans from other primates—bipedality, manual dexterity, elaborate tool making, intergenerational transfer of technical skills, and higher encephalization—was fostered by our evolution from simple foragers to sophisticated *hunters* and incipient plant cultivators.

The earliest foragers were almost certainly opportunistic scavengers, taking advantage of partially eaten herbivore carcasses left behind by large predators—or at least breaking the bones to extract the nutritious marrow. They were obviously omnivorous, collecting and killing scores

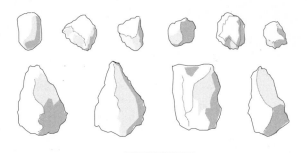

Comparison of the oldest Oldowan tools with larger Acheulean hand axes used to butcher animals.

of different edibles, but a small number of foods was usually dominant. Large roots—easily found by associated leaves or stalks and dug out quickly with the help of pointed sticks—provided the highest (up to fortyfold) net energy returns in gathering. But unlike *grains,* which were also easy to collect and had higher energy content, they were low in protein. High returns in collecting large seeds and nuts were reduced by often considerable energy expenditures in their processing.

Typical gathering of a wide variety of plant foods returned at least ten to fifteen times the invested energy, ratios similar to killing large mammals—while hunting smaller animals yielded much smaller energy returns. Foraging was especially unrewarding in tropical *forests* where edible fruits and seeds are a very small share of total plant mass and are mostly inaccessible in high canopies. And because most tropical mammalian *herbivores* are arboreal, their hunting also yielded low energy returns. This explains why there is no unambiguous ethnographic accounts of tropical foragers who would not engage in some plant cultivation.

In contrast, *grasslands* provided excellent foraging environment. Grass seeds and starchy roots were easy to collect, and there were many large *herbivores.* Cooperative hunting of large ungulates—much more rewarding than solitary pursuits—clearly made lasting contributions to human socialization. Many large mammals could be killed by skilfully driving them into confined runs and capturing them in pens or natural traps, or by stampeding them over cliffs.

But simple energy ratios favoring the killing of bison over the snaring of hares does not capture the desired quality of hunted food. Foraging diets were often low in *lipids,* and inland *hunters* could satisfy the apparently universal craving to eat filling fat, rather than just lean *meat,* only by killing larger fatty mammals. This is why the African *hunters,* exploiting the unique human capacity for long-distance *running,* were willing to chase big antelopes to exhaustion; this is why their counterparts in boreal Europe were willing to face huge woolly mammoth with their simple spears; or why North American Indians invested so much effort into driving bisons across precipices: energy return on their *labor* investment was eating food uncommonly filling with *lipids.*

Maritime *hunters* could get similar rewards with usually much less exertion. Killing of migrating baleen whales by the Northwestern Alaskan Inuit provided more than a

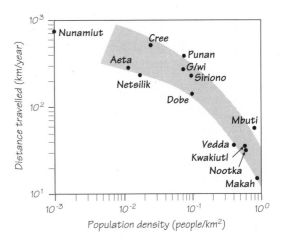

Relationship between population density and annual travel in foraging societies: Maritime cultures could be least mobile.

two thousand-fold net energy return, perhaps the highest documented foraging gain. Much lower, but still very comfortable, ratios were obtained by killing seals or catching migrating salmon. Seasonal abundance of such foods could be also easily preserved by drying or smoking and stored for later use. Large-scale food storage helped to stabilize populations at higher densities: maritime foragers could cease roaming and could live in fairly large, permanent settlements with social stratification, elaborate rituals, and long-distance trade.

Except for some maritime cultures, foraging societies could not attain population densities needed for functional and social diversification. The least hospitable environments (tundras, boreal *forests*) supported population densities of just 1 person/km², while the most suitable habitats (tropical and temperate *grasslands*) could carry from ten to a hundred times more people. Because of differences in phytomass storage, accessibility and edibility of plant parts, and sizes and habits of hunted animals these large density variations had no simple correlations with total *photosynthesis* or biodiversity of habitats.

Approximate contributions of gathering, hunting, and fishing to typical diets of some foraging societies surviving into the twentieth century.

Burning of forest phytomass, the first
step in creating temporary fields.

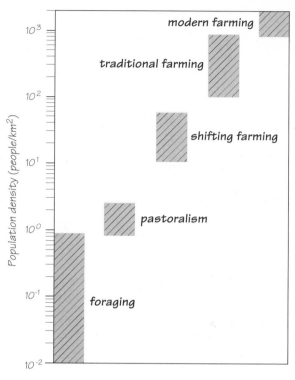

Shifting cultivation could support population densities two
orders of magnitude higher than foraging—but an order of
magnitude lower than traditional farming.

Although some foraging societies had to spend just a few hours a day to enjoy abundant food supply, other had to cope with recurrent hardships intensified by seasonal food shortages leading to high infant mortalities (and to infanticide) and to often devastating famines. The idea of foraging existence as the original affluent society is clearly an impermissible generalization.

Shifting Cultivation

Adoption of plant cultivation, and in most cultures also of associated domestication of animals, was a multifocal and gradual process stimulated by population growth beyond the densities supportable by foraging and by environmental changes (such as a drier climate or elimination of previously abundant prey) reducing food supply that *hunters and gatherers* could get by foraging. Diminishing energy returns in foraging favored more regular reliance on incipient cultivation begun with the deliberate planting of tubers or scattering of seeds that was practiced by many *gatherers*. Domestication of wild *grains* and *cattle* offered a supply of high-quality *nutrients*.

Overall energy returns of crop cultivation were commonly lower than those for many kinds of foraging, and that is why *hunters and gatherers* coexisted with settled agricultural societies for hundreds, and sometimes even for thousands, of years. But even extensive cultivation could support higher population densities and assure a more reliable food supply than that enjoyed by most foraging societies. The fastest transition to permanent cropping happened on fertile floodplains, but elsewhere *agriculture* began, and continued for often very long periods, as shifting cultivation.

This practice extended from rainy tropics to the subarctic forests, and it encompassed a wide variety of species and local peculiarities. But everywhere it alternated between short periods of cropping (commonly just one season, rarely more than three years) and long spans of fallow (sometimes up to twenty-five to thirty years, more commonly at least a decade). The cropping cycle began with often only partial removal of natural phytomass. In *forests* and shrublands this was done by a combination of felling, slashing, and burning; in *grasslands* just by setting fires. Nearly all *nitrogen* was lost in combustion, but mineral nutrients helped to produce at least one or two fairly good crops.

Cultivated species included cereal and leguminous grains, tubers (sweet potatoes, cassava, yams), vegetables, fruits, fibers, and medicinal plants. Few staples provided most of the food nutrients, but the total number of planted species was rarely less than a dozen, and in warmer environments it often surpassed two scores. Cultivation was done commonly in jumbled, gardenlike arrangements, with high degrees of interplanting and intercropping. Net energy returns for grains, be they tropical (dryland Asian rices), subtropical (African millets), or temperate (European rye) ranged mostly between ten and fifteen. Less labor was needed for corn (its energy returns were commonly more than twentyfold) and for tropical tubers, legumes, and bananas (the best energy returns ranged between forty and seventy). Ancient Mesoamerica, with staples of corn, beans, and squash, achieved some of the highest food production efficiencies, but even less efficient shifting cultivation could support population densities ten times higher than those of prosperous *hunters and gatherers*. Minima were about 0.2 people per hectare in Eastern North America (corn gardening), maxima surpassed 0.6 people per hectare in Southeast Asia (upland rice and roots). High energy returns sustained *shifting cultivation* in parts of Asia, Africa, and Latin America into the twentieth century. Higher population densities and shorter regeneration cycles have been the main reasons for its recent decline.

Traditional Agricultures

Differences in climate, soils, crops and in specific farming practices created a great variety of *traditional agricultures*—but physical imperatives of field cultivation dictated a recurrent pattern of *labor*. Throughout the Old World the sequence begun with plowing and harrowing which prepared loosened, weed-free, well-aerated, and leveled soil ready for sowing (mostly by hand). Plow designs progressed from the simplest symmetrical scratch plows, which created merely a shallow furrow for seeds, to asymmetrical moldboard plows—first just straight pieces of wood, later with curved metal plowshares—which turned soil over, buried the cut weeds, and eliminated the need for crossplowing. Harvesting was done, slowly, by sickles, later by more efficient scythes (the first successful mechanical reapers were introduced only after 1830).

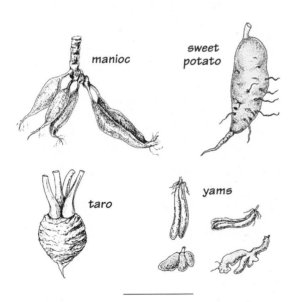

Roots have been dominant staples in tropical shifting cultivation.

Mowing with scythe and binding and stooking sheaves in sixteenth-century Netherlands (Peter Brueghel the Elder's *Summer,* 1565).

tion and fertilization, which improved the supply of two inputs most often limiting crop productivity; water and plant nutrients; and on cultivation of a greater variety of plants, which made cropping not only more productive but also more resilient.

Deeper plowing in heavier soils was the most indispensable service provided by harnessed **cattle** or **horses.** Animals were also used commonly for threshing and oil extraction, and to lift water for irrigation. But even in relatively well-off European countries human exertions—commonly requiring a great deal of heavy or even very heavy **labor**—remained dominant in sowing, hoeing, weeding, transplanting, and harvesting until the nine-

Cropping in all of the Old World's cultures was dominated by cereal **grains:** millets and rice in China, wheat and rice in India, wheat and barley in the Middle East, sorghums in Sahelian Africa, and wheat, barley rye, and oats in Europe. Plowless Mesoamerican societies relied on corn, and even the Incas probably derived as much food energy from quinoa and corn as they did from potatoes, if not more. Generally low yields were further reduced by large storage losses and by relatively high seed requirements.

The earliest cropping could support, with just a single harvest of a staple crop yielding no more than one tonne per hectare, about ten times as many people as the same area would in **shifting agriculture.** Higher harvests were achieved either by extension of arable land, or by intensified cropping. Gradual intensification rested on three essential advances: on greater use of draft animals, which eliminated much of the heaviest human **labor;** on irriga-

As François Millet's *Sower* (drawn in 1825) shows, grain sowing was done by hand even in nineteenth-century Europe.

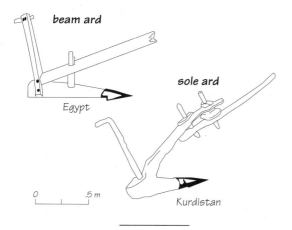

Two kinds of primitive scratch plows from Egypt and Kurdistan.

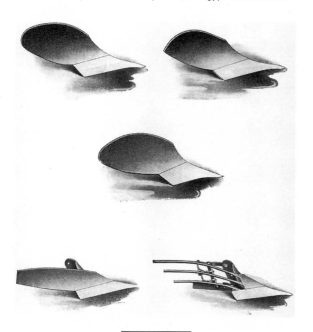

Different kinds of curved steel plowshare produced
in the United States during the 1880s.

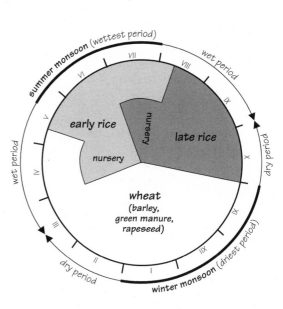

Cropping calendar for intensive rotations
in South China.

teenth century. Harvesting of *grains* was usually the most demanding task: it took three to four times longer than plowing. A disproportionate share of total farm *labor* was done frequently by women, and even very young children helped in fields, gardens, or yards.

Irrigation relied on a variety of mechanical devices lifting water from adjoining canals or from wells. They ranged from simple counterpoise lifts and ingenious Archimedean screws and wooden ladders to large wheels with clay pots or bamboo tubes fastened to wheels. Higher yields required higher nutrient inputs, and recycling of various organic wastes eventually surpassed ten tonnes per hectare in the Old World's most intensive *agricultures.* Planting of legumes, as food, feed, or green manures, was another important source of *nitrogen.*

Both extended cultivated and intensified cropping needed higher investment of *labor,* even in the regions

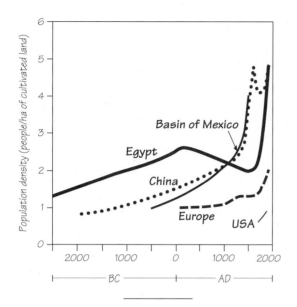

Long-term increases in population densities
per hectare of arable land.

where abundant arable land could produce plenty of feed for more draft animals. Intensified cropping provided almost always lower energy returns than extensive cultivation or *shifting agriculture*. That is why peasants expanded or intensified their cropping only in response to gradually growing populations. This reluctant expansion and intensification was eventually able to support impressively high population densities. In terms of national, or large regional means, the ratio reached maxima of three to four people per hectare in temperate environments, and up to five to six people per hectare in monsoonal climates. Peaks in the most intensively farmed areas (Japan's Kanto Plain, China's Sichuan) were up to twelve to fifteen people per hectare with overwhelmingly vegetarian diets. This limit to traditional farming was imposed by the availability of recyclable *nitrogen*.

But these accomplishments were not generally accompanied by higher per capita food availabilities (as even coarse *bread* was in short supply in parts of the early nine-teenth century Europe) or by a better quality of average diets (in Europe they remained monotonous until the nineteenth century, in Asia well into the twentieth). No traditional agriculture could also produce enough food to eliminate extensive malnutrition. Peasant reluctance to increase *labor* inputs in more extensive or more intensive cropping together with the preference for large families minimizing individual work loads had made it difficult to raise average per capita food supply and to avoid recurrent food shortages. Only rising inputs of *fossil fuels* into cropping — directly to power machinery, and indirectly to produce it and to synthesize chemicals — could support both larger population and higher, and better, average food supply.

Cattle

Rise of the first civilizations was closely connected with the domestication of *cattle*. Harnessing of oxen, cows, and water buffaloes made it possible to begin cultivation of larger fields: even when harnessed to a primitive plow with a wooden share, an animal would cultivate a field at least three times faster than a peasant with a hoe. Plowing and *cattle* domestication coincide in all Old World societies, where bovines became eventually the dominant source of animate power.

For millennia, their slow pace determined the tempo of everyday rural life. They provided inexpensive and indispensable draft not only for all field tasks but also for raising water, processing crops (from primitive grain threshing by trampling to turning oil seed and sugar cane mills), and transporting people and goods. In Europe cattle's primacy was displaced only gradually by *horses*, and several hundred million head of cattle are still working in nearly a hundred poor countries during the last decade of the twentieth century.

This persistence cannot be explained by superior performance. Lower weights and slower gaits made harnessed bovines almost always less powerful draft animals than *horses*. Because sustainable tractive force is roughly proportional to body mass even average oxen had a 15–25 percent disadvantage compared to good *horse* breeds, and their lower speeds—usually no more than two-thirds of the steady *horse* pace—made it impossible to develop more than three hundred to four hundred watts of draft power during steady work, compared to five hundred to six hundred watts even for relatively weak *horses*.

Naturally, for shorter periods the exertions can be much higher. Maximum measured two-hour pulls for pairs of good German mountain cows were nearly identical to the performance of small *horses,* but when comparing averages the advantage remains with *horses*. Even when working in pairs, *cattle* may have a difficult time with such common field tasks as deep plowing (which needs pulls of 120–170 kg), heavy harrowing (pulls com-

Water buffaloes milling rice.

Holstein

Madagascar

Blanco orejinegro

Huangniu
(Chinese yellow
cattle)

Different breeds of cattle used both as dairy and draft animals.

monly around 100 kg), and grain harvesting with a mechanical reaper and binder (pulls of up to 200 kg).

Nor is it possible to harness cattle more efficiently than can be done with collared *horses*. Various head and neck yokes are inherently more awkward and less comfortable. The Mesopotamian double head yoke, the most ancient harness fixed either at the front or the back of the head, is unsuitable for long-necked animals. Throat fastenings of the double neck yoke tend to choke the animals who must be of the same size to prevent even more individual discomfort.

Explanation of cattle's traditional usefulness lies above all in their energetic advantage: as ruminants they did not create any additional demand for feed *grains* grown on good farmland. They can extract sufficient nutrition just from *grasslands* and crop residues, with occasional supplementation by such crop processing by-products as milling, or oilseed-pressing wastes. Even

Oxen with a double-head yoke.

densely populated Asian lowlands with extremely limited grazing could thus support large counts of working bovines without reducing food production capacity.

In some regions there was virtually no competition between cattle and people for land, carbohydrates or protein, in others fodder crops for cattle took up only a few percent of all cultivated land. In contrast, as *horses* got heavier and worked harder, they needed cereal or legume *grains* whose cultivation claimed increasing shares of cropland.

Much less important is cattle's slight mechanical advantage. Everything else being equal, efficiency of field work will be higher with smaller bodies because the line of pull deviates less from the direction of traction. Moreover, the less angled pull also reduces the uplift on implements: plowing with small *cattle* will put much less strain on a plowman than working with large *horses.*

And, of course, *cattle* were also a source of fertilizer needed for intensive cropping and providers of high-quality protein. Healthy ruminants produce at least 50 percent more waste per unit of body weight than *horses,* but the *nitrogen* content of freshly voided bovine and equine manures is very similar. In many regions where traditional diets were almost totally vegetarian — by choice or

due to high population densities — dairy products provided the only regular source of animal protein.

Bovines can also work well in environments where *horses* never could. Water buffaloes are particularly good example of such a fit. Male castrates used for work are as heavy as good-size *horses,* and in spite of their clumsy look they move easily on narrow field dividers or in muddy fields, helped by their large hoofs and flexible pastern and fetlock joints. Their aquatic adaptation allows them to graze while completely submerged, tapping phytomass inaccessible to any other working animal.

Horses

In big cities *horses* were displaced by streetcars and automobiles during the early decades of the twentieth century, but their draft had dominated even the rich world's farming until the 1940s. They used to come in many varieties and sizes, from light (less than 400 kg) Asian breeds to heavy

Silhouettes of two small and two large European draft horses.

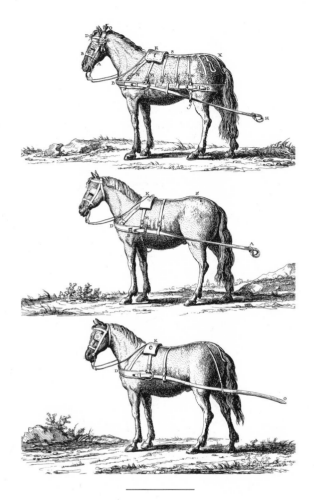

Breastband harness from nineteenth-century France.

useful work at a rate of about 750 watts, or one horsepower.

All of these power ratings were about double the values for similarly sized *cattle*. The superiority of nineteenth-century horsedrawn field cultivation was obvious: opening of North American plains could not have been accomplished so rapidly with oxen. Advantages did not end with higher power outputs. Well-fed *horses* also had greater endurance, and during brief exertions they could pull up to 35 percent of their body weight, an equivalent to about three horsepower!

And *horses* also require hardly any additional energy for standing. While in other mammals standing claims at least 10–15 percent more energy than lying, a *horse's* unusually strong suspensory and check ligaments mean that the animal can rest comfortably, and even doze, while standing in harness, or it can graze leisurely with only a negligible energy markup above its basal *metabolism.*

But harnessing *horse* power posed historically two difficult challenges, one mechanical, the other metabolic.

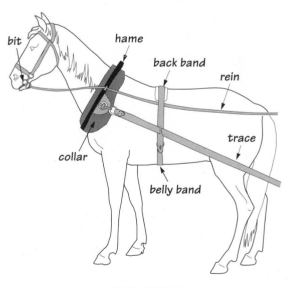

Collar harness of a twentieth-century draft horse.

chestnut Belgians, black or grey Percherons of the French Northwest, or Clydesdale bays or browns of Scotland.

The heaviest of these animals approached, or even topped, one tonne. Prolonged pulls for all breeds were about 15 percent of body weight and typical speeds of around one meter per second combined to deliver draft ranging from 500 to nearly 1,500 watts of sustainable power. A healthy five hundred-kilogram *horse* delivered

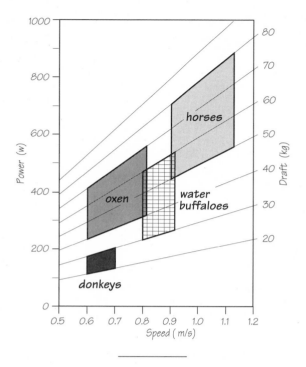

Comparisons of speed, draft and power for
common working animals.

All ancient horse-using cultures had throat-and-girth harness, but this arrangement was good for only relatively light pulls. Its point of traction was too high, precluding the optimal transfer of power: the lower and the more parallel the line of pull is with the direction of traction the higher the efficiency of work. Moreover, the throat strap, used to prevent the backward slippage of the girth, was depressing the trachea and choking the animal during heavy exertion when it lowered and advanced its head. The early medieval breastband harness removed this constraint, but its point of traction was far away from the animal's powerful pectoral muscles.

Only the collar harness, originating in northern China during the Han dynasty and adopted in Europe by the ninth century, allowed for the most comfortable and the most efficient draft: it consisted of a single oval wooden frame that was lined for a comfortable fit onto a *horse*'s shoulders, with traces connected to the hame just above the animal's shoulder blades. In comparison to a team in a choking harness, a pair of animals with comfortable collars could pull loads at least four and up to ten times heavier than they could with throat-and-girth harness. After the ninth century the *horse*'s performance was further enhanced by the general adoption of iron horseshoes, preventing excessive wear of soft hooves and improving the animal's traction and endurance. Swingletrees attached to traces equalized strain resulting from uneven pulling and enabled plowmen to harness any even or odd number of *horses*.

Combination of the collar harness, iron horseshoes, and swingletrees made it possible to clear dense *forests* by pulling away the cut logs and wrenching out the stumps, to plow heavy forest soils into new fields, to haul massive stones from quarries to cathedral sites, to move big road wagons supplying growing cities and marching armies. With such innovations the medieval world was changed.

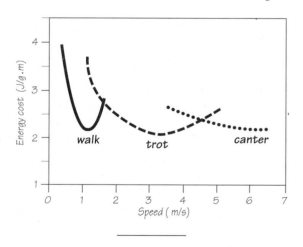

Energy costs of horse gaits.

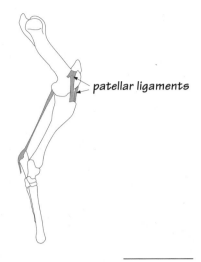

patellar ligaments

The stay mechanism of horse legs: Patellar ligaments lock one of the four angles of a parallelogram.

But these exertions had to be energized by better feed than just roughages—*grasses* or *straws*—sufficient for working cattle. Powerful *horses* also had to be fed some concentrates, cereals, or legume *grains* with higher protein content. Intensive cropping started its gradual diffusion as medieval peasants had to change their traditional practices in order to provide for both their families and their animals. During the late nineteenth century American *horses* were fed four to five kilograms of oats a day, and at the peak of their importance, between 1910 and 1920, at least one-fifth of U.S. farmland had to be devoted to cultivation of *horse* feed!

One well-fed American *horse* preempted cultivation of food *grains* capable to sustain about six people—but it could work at a rate at least ten times higher than an average man, offering a substantial energy advantage. But new farm machines required a more concentrated source of power: it was a logistic mess to harness and to guide two or even three dozens of *horses* pulling grain combines on

California's vast grainfields. Even the early, relatively small, *internal combustion engines* mounted on tractors or used for pumping water or in threshing, could replace at least ten *horses*—and do so while claiming no land for feed. The millennium of *horse* power that built Western civilization came to a rapid end.

Wood, Charcoal, and Straw
Once accepted, even misleading labels tend to stick. In 1836 Christian Thomsen, a Danish historian, chose to divide human evolution into the Stone, Bronze, and Iron Ages. Although it is not inaccurate in describing the

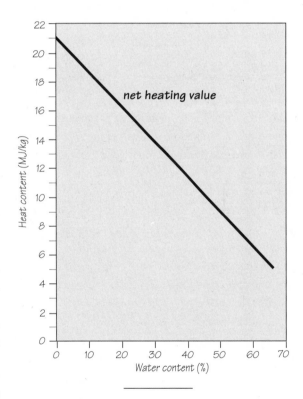

The net heating value of wood declines linearly with rising water content; wood with more than 67 percent moisture cannot be ignited.

Douglas fir from the Cascade Mountains cut at twenty-five years of age (actual diameter is about 23 cm).

climates air-dried *wood* still contained about 15 percent of moisture.

Charcoal contains a mere trace of moisture and its energy density is comparable to that of very good *coals*—but because it is virtually pure carbon its burning is smokeless, a great advantage for indoor uses. *Charcoal* is also excellent for firing bricks, tiles, and lime, and for smelting ores. Its high porosity eased the ascent of reducing gases in *blast furnaces,* but its friability limited their height to less than eight meters. Unfortunately, traditional charcoaling was very wasteful: the product yield in primitive kilns was just 15–25 percent of the charged air-dry *wood,* for the loss of about 60 percent of the original energy.

Dependence on wood and *charcoal* put little appreciated limits on the size of ancient cities. Because *photosynthesis* converts typically less than one percent of *solar radiation* into new phytomass, the best annual productivities of fast-growing trees, even in warm and rainy climates, are rarely above fifteen tonnes per hectare; in drier regions the best rates are between five and ten tonnes per hectare. With average heat value of 18 MJ/kg of dry *wood,* these yields imply power densities of 0.3–0.9 W/m², and inefficient conversion of much of this *wood* into *charcoal* would lower these values to between 0.2 and 0.5 W/m².

In contrast, a large Wooden Age city in a colder climate (in Northern Europe or in North China) would have consumed at least 20–30 W/m² of its built-up area, mainly for heating and cooking, and also for manufactures ranging from blacksmithing to firing of tiles. Consequently, power density of sustainable *forest* growth in temperate climates was commonly equal to less than one and rarely more than 2 percent of the power density of urban energy consumption—and the cities required nearby areas anywhere between fifty to two hundred times their size to satisfy their thermal energy needs. This necessity limited their population size and manufacturing intensity even if food and water were abundant.

choice of materials for a number of essential tools and weapons, this classification exaggerated the importance of metals in everyday life of most of the people, that is poor peasants, in preindustrial societies. Those who lived in or near *forests* lived for millennia in the Wooden Age. Aside from clothes and a few tools nearly all of their meager possessions were wooden: their houses, furniture, bowls, buckets, spoons, plows, and harrows. And, of course, they cooked and heated their dwellings with *wood,* and they used it to make *charcoal* for cleaner heating and cooking, as well as for smelting and refining metals.

Households collected *wood* often as fallen, broken, or lopped-off branches, twigs, bark, and dead roots: Even small children could help, and the dry *wood* burned well. In contrast, harvesting stem *wood* and heavy branches required good axes or saws, it had to be dried before combustion because it does not ignite when it has more than 67 percent of moisture, and it burned very inefficiently when the moisture was above 40 percent. But even in dry

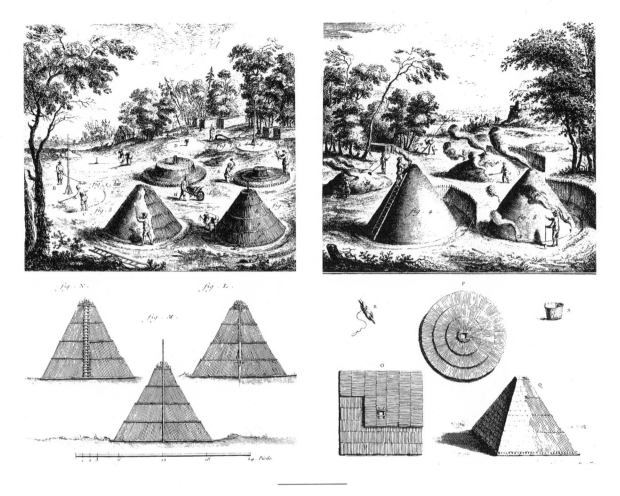

Charcoal making in the eighteenth century.

Annual per capita *wood* requirements were well below one tonne only in the poorest villages of tropical regions. Two tonnes were needed in climates with pronounced winters and with substantial smelting of metals. European and North America needs at the very end of the Wooden Age (during the eighteenth and nineteenth centuries) were higher still, reflecting large consumption of *charcoal* by *blast furnaces* and in firing bricks and tiles. In colder regions consumption ranged up to six tonnes, and in the United States even in 1850 the nationwide average was about five tonnes.

On deforested, intensively cultivated agricultural plains and in arid, sparsely treed areas crop residues, most commonly cereal *straws* and stalks, supplied much of the needed fuel. Ripe *straws* have very low moisture content, but their low density made for bulky storage and inconvenient burning: fires had to be restoked very frequently. And because of competing demands — as sources

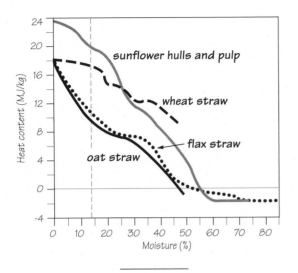

The net heating content of some crop residues.

of recycled organic *nitrogen,* as animal feed and bedding, as materials for thatching and small manufactures — crop residues were often in short supply.

Yet another biomass fuel, dried animal dung, has been of a great importance in many traditional societies. Llama dung was burned on the Andean Altiplano; cattle (and also camel) dung was always used in the Sahelian Africa, in Egyptian villages, and in Asia's arid interior (from Afghanistan to Mongolia), as well as in its monsoonal regions; and yaks provided fuel on the high-lying Tibetan plateau. Wild buffalo and cattle dung also made possible the first continental crossings and the subsequent colonization of the Great Plains during the nineteenth century: Early settlers in treeless landscapes depended on this fuel to survive bitterly cold winters.

Waterwheels

We are not certain about either the time or the place of origin of first *waterwheels.* The indisputable history of converting the kinetic energy of streams into the rotary motion of millstones starts with first Greek and then Ro-

man references during the first century B.C. The two distinct designs — simple horizontal and more complicated vertical arrangements — emerged almost concurrently, but their subsequent diffusion, increase in power and improvements in efficiency were very slow.

A horizontal wheel was simple to make: paddles were attached to the lower end of a sturdy wooden shaft whose upper end was inserted into a millstone. No gearing was required, but the efficiency of these usually fairly crudely made machines was low, and their uses were limited

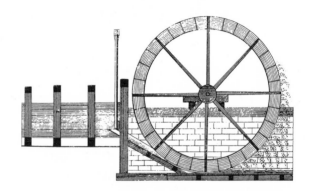

Section and a view of two eighteenth-century undershot wheels.

Section and a view of two eighteenth-century overshot wheels.

Their designs made it possible to harness different kinds of streams. Undershot wheels did not require any significant differences in water level; they were put directly into a stream to be turned by its kinetic energy. Because the power of a moving medium increases with the cube of its speed, undershots worked best in fast-flowing waters. And because frequent and pronounced water-level fluctuations would leave them either swamped or dry, they were best suited for streams with fairly steady flows. First undershots were built with simple radial boards, and later they were fitted with backs that prevented water from shooting over the floats. Their most efficient design, the Poncelet wheel with curved blades, was introduced only at the beginning of the nineteenth century.

Breast wheels were driven by a combination of kinetic and gravitational energies. They required water heads of at least 1.5–2 meters, and if they were to work efficiently they also had to have fairly close-fitting stone or wooden breastworks, preventing premature spilling. In overshot wheels gravity dominated the energy conversion.

mostly to small-scale milling of *grains.* Still, this simple design persisted throughout the Middle Ages, and a vastly more efficient reincarnation of the horizontal wheel concept came only during the nineteenth century with the development of first *water turbines.* In contrast, vertical *waterwheels,* inherently more powerful and more efficient, became the leading inanimate prime movers of the preindustrial West.

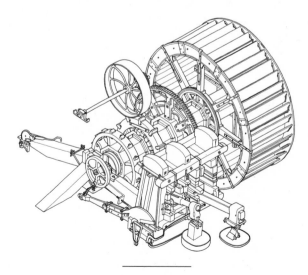

A modern drawing of a large nineteenth-century overshot waterwheel powering forge hammers.

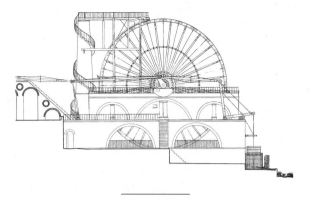

Section of *Isabella*, the largest waterwheel, completed in 1845.

Consequently, they required water heads of at least three meters, but as long as the flow was steady it could be slow. Various improvements assured the combination of high head and steady input. Impoundments—ranging from low weirs and small dams to deep ponds—reduced fluctuations in water supply and often also increased the available head, a goal accomplished commonly by digging contour-following canals or constructing elevated races.

Naturally, all vertical wheels needed gears to convert rotary motion to useful work, and all early machines lost a great deal of their power in this process. Eventually, standard overshot efficiencies rose to more than 60 percent, and the wheels did not only do most of the preindustrial grain milling, but also increasing shares of a widening variety of mining and manufacturing tasks ranging from water pumping from deep shafts to wire pulling. Even the earliest *waterwheels* had a considerably higher power rating than all contemporary animate prime movers, and a slow, but steady, rise of capacities eventually made the machines indispensable energizers of the early stages of industrialization.

While human *labor* could deliver around one hundred watts of sustained power per person, and *cattle* usually no more than three hundred watts per head, typical Roman *waterwheels* rated around two kilowatts. Corre-

sponding increases in productivity were impressive: a slave could grind about three kilograms of grain per hour, a donkey about ten, but waterwheel-driven millstones could grind up to one hundred kilograms. Typical unit sizes of vertical wheels surpassed four kilowatts only late in the eighteenth century, but then they multiplied rapidly as the machines reached the peak of their importance. Wheels with capacities up to fifty kilowatts—more powerful than many contemporary *steam engines*—became common by 1830, and their groupings along steep conduits fed from reservoirs resulted in installed system capacities surpassing one megawatt.

The largest waterwheel ever built was not sited either on a major river or on a rushing mountain stream, but on the bottom a hill slope on the Isle of Man. Water from all the streams above the wheel was led from collecting tanks to the base of a masonry tower, and from there it rose to a wooden flume and fell on the nearest waterwheel quadrant, making *Lady Isabella* a pitchback overshot. The wheel was large (diameter of 21.9 m) but narrow (only 1.85 m, compared to up to 6 m for other large machines), and although it turned slowly, just 2.5 times a minute, it transmitted about two hundred kilowatts of useful power to the pump at the bottom of a 450-meter shaft of the lead-zinc mine of the Great Laxey Mining Company. Finished in 1854, the machine was not a harbinger of greater things to come: The rapid rise of *steam engines* and *water turbines* made the *waterwheels* obsolete just a few decades after they reached their apex.

Windmills

Like *waterwheels, windmills* can be horizontal or vertical, and the first wind-driven rotaries were clearly reminiscent of their older water-driven counterparts. These simplest horizontal *windmills* had rectangular cloth-covered sails mounted on vertical wooden shafts directly turning the millstones. Al-Masudi saw them before the middle of the

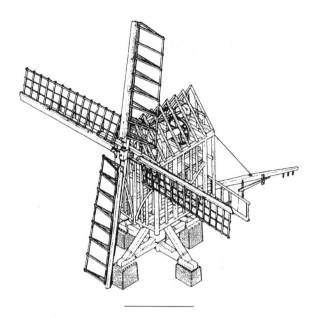

A sturdy medieval post mill built on four massive quarterbars.

tenth century in Seistan, a windy region of today's eastern Iran: "Seistan is the land of winds and sand. There the wind drives mills and raises water from streams, whereby gardens are irrigated. There is in the world (and God alone knows it) nowhere where more frequent use is made of the winds."

How old this practice was remains unclear, but we know that Seistan's unique *windmills* did not diffuse widely. Again, much like in the case of *waterwheels*, the wind-powered machines that made the greatest contribution to preindustrial societies were vertical, mounted on horizontal shafts, and hence required gears to transmit their rotation to useful work. By the second half of the twelfth century they became common in those parts of Europe that had few rushing *rivers* but strong prevailing *winds*: throughout northern France, in Flanders, and in southern England. From there they diffused eastward and southward, but parts of the Iberian peninsula retained oc-

tagonal sail mills with triangular cloth, which originated in the eastern Mediterranean.

Medieval vertical *windmills* pivoted on a single massive wooden post supported by diagonal quarterbars. The whole pivoting structures of these post mills, housing gears and millstones, had to be turned manually, and laboriously, into the wind, and realigned once its direction had shifted. They were frequently toppled by strong *winds,*

Cross-section of a large eighteenth-century French post mill showing the gears and millstones.

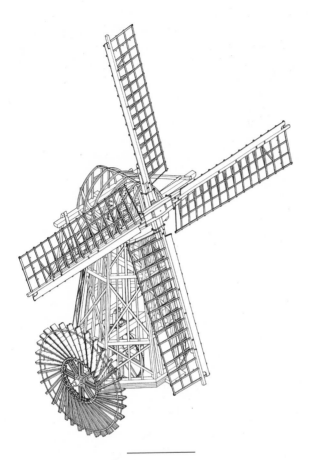

A large Dutch tower mill powering a drainage wheel.

sisted of a fixed body surmounted by a rotating top, which was much easier to turn into the wind. In the earliest designs this was done from the ground; later it could be done more conveniently from circular galleries, and after 1745 the invention of fantail powering a winding gear made it possible to keep the sails automatically into the wind. And, of course, the greater height of towers made it possible to harness faster and hence more powerful *winds.* Tower, or smock, *windmills* reached their greatest importance during the eighteenth and nineteenth centuries in Northwestern Europe, in a roughly elliptical region whose long axis extended from Normandy to Denmark, and, with the exception of mining operations, they were eventually used for as many different tasks as *waterwheels.* Per-

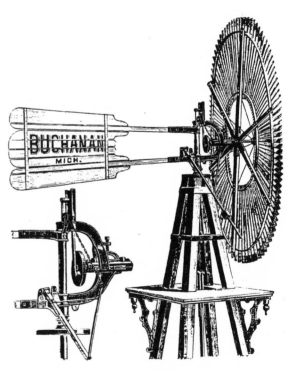

Multivane windmills dominated American market of the late nineteenth century.

and because their sails were close to the ground their maximum power was limited. The theoretical capacity of *windmills* increases with the cube of the wind speed, which in turn is proportional to the height above the ground raised to the power of 0.14. Consequently, for a given wind speed near the surface, a machine with a shaft ten meters above the ground will be only about 60 percent as powerful as one with its axle thirty meters above the ground.

Tower mills eventually solved all of the shortcomings of medieval post mills. Their massive structures built of stone, bricks, or wood provided great stability. They con-

haps their most outstanding historic contribution was to power wooden wheels with scoops and Archimedean screws that drained large areas of the Netherlands.

But the most numerous *windmills* of the nineteenth century were of a very different design: they served small farms and railway stations on the windy Great Plains during the period of rapid westward expansion of America's population. These compact and efficient machines did not have large sails but numerous narrow slats fastened to solid or sectional wheels on top of simple, cross-beam stiffened, wooden towers. Scores of manufacturers turned out several million such machines during the latter half of the nineteenth century. Together with the Colt revolver and barbed wire they opened up the Plains for settlement.

Until the early nineteenth century *windmills* in common use were roughly as powerful as their contemporary water-driven mills. We have no reliable estimates for earlier centuries, but after 1700 European post mills rated mostly between 1.5 and 6 kilowatts, and tower mills between 5 and 10 kilowatts in terms of useful power. This power is necessarily only a fraction of the available flux: If all of it were converted to kinetic energy of rotation the airstream would have to stop and the air would accumulate at the point of conversion.

The maximum convertible share, the so-called Betz limit, is 16/27 of the flow, or almost 60 percent of the kinetic energy in the wind. Modern airfoils extract about 50 percent of the available flow, and typical performances for preindustrial *windmills* were between 20 and 30 percent. Consequently, the largest eighteenth-century mills harnessed about thirty kilowatts of useful power, and improved sails and metal gears could raise their output to around forty kilowatts during the early 1800s. But by that time many *waterwheels* had capacities two to three times larger. In contrast, small American *windmills* rated just 30–150 watts for common farmstead machines, and around one kilowatt for larger commercial designs.

Forging metal in an ancient Greek workshop.

Copper, Iron, and Steel

Traditional metallurgy began with *copper* sometime after the middle of the fourth millennium B.C. in a number of Middle Eastern locations. From there the smelting diffused both to the Western Mediterranean and to Central Asia. *Copper*'s rather high melting point (1083°C), and the multistep smelting process required for producing the metal from abundant sulfide ores, added up to large *wood and charcoal* needs. Ores had to be first crushed and roasted in order to remove sulfur and other impurities (Sb, As, Fe, Pb); until *waterwheels,* or *horses* harnessed to whims, took over the task, ore crushing was done usually by laborious hand-hammering.

Smelting of roasted ores—first just in clay-lined pits, and eventually in clay shaft furnaces with bellows—was followed by smelting of the coarse metal and by its re-smelting yielding nearly pure (95–97 percent) blister *copper* that could be refined even further. These practices remained largely unchanged until the early modern era, and their high energy requirements, exacerbated by inefficient firing of *charcoal,* resulted in extensive deforestation in

Casting bronze bells in mid-eighteenth-century France.

many ore-rich regions from the Iberian peninsula to Afghanistan.

Copper was commonly alloyed with tin, whose low melting point (231.97°C) needed little *charcoal* for the metal's reduction from crushed oxide ores. Resulting bronzes, containing 5–30 percent tin, were thus less energy-intensive than *copper,* but both their tensile strength and hardness were at least 2.5 times higher than for pure *copper,* making it possible to turn out first good metallic tools and weapons—and bells (their bronze had 25 percent tin). Brass, combining *copper* with zinc, was a much less common ancient alloy.

Substitution of *copper* by *iron* was a long-drawn process. In Mesopotamia the metal became common after 1000 B.C., in China some four centuries later, but it was never smelted by any New World society. Because *iron* melts at 1535°C its smelting required *charcoal* burned with forced air supply: such combustion can produce temperatures close to 2000°C, compared to just 900°C for unaided fire. *Iron* smelting began in shallow, clay- or stone-lined pits often located on hilltops to maximize natural draft. Later, clay shafts were built, and insertion of a few narrow clay tubes (tuyeres) connected to bellows delivered oxygenating blast into the hearth. This smelting could not produce liquid *iron,* only a spongy mass of *iron* globules, slag, and cinders.

Reheating and hammering of this mass yielded wrought iron, a tough and malleable metal with a mere trace of carbon. Han dynasty (207 B.C.–A.D. 220) Chinese were the first producers of liquid *iron* thanks to ores high

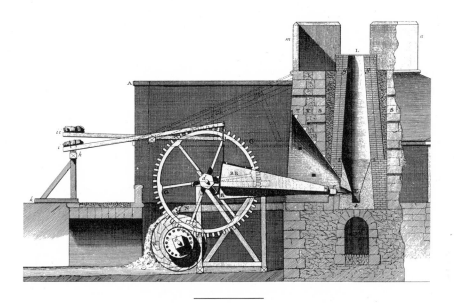

Cross-section of a charcoal-fueled iron blast furnace
with bellows powered by an overshot waterwheel.

in phosphorus, which lowered iron's melting point, as well as to the deployment of double-acting bellows that provided a strong air blast. But the origins of modern *blast furnaces* are most likely in the lower Rhine valley around the year 1400. As their stone or brick stacks grew wider and taller, they needed more powerful bellows operated by *waterwheels.* By the late seventeenth century waterwheel capacities became a limiting factor in *iron* smelting, as did *charcoal*'s inability to support more massive ore and limestone charges. Only the invention of more efficient *steam engines* and production of coke from suitable *coals* did away with these limits.

Advances in traditional ferrous metallurgy are best appreciated by comparing typical energy needs. Medieval hearths needed four to eight times more fuel than the mass of ore. Even with good ores this meant at least ten (and up to twenty) kilograms of *charcoal* per kilogram of hot metal. By 1800 the typical *charcoal*/metal ratio was down to about 8, by 1900 to just 1. 2. The high *wood* demand of medieval and early modern *iron* smelting created many deforested landscapes. England's early adoption of coke is easy to understand: A single early eighteenth-century furnace consumed annually a circle of *forest* with a radius of about four kilometers. And if American ironmakers had not switched to coke after 1870, by 1900 they would have consumed annually enough *forest* to fill a square whose side would be the distance between Boston and Philadelphia.

Blast furnaces produce cast (pig) *iron,* whose high (1.5–5 percent) carbon content makes it hard but brittle: the metal cannot be directly forged or rolled. Decarburization of pig *iron* (by oxygenation), or carburization of wrought iron (by heating it in *charcoal*), produced *steel,* containing 0.15–1.5 percent carbon, whose tensile strength and hardness are superior both to cast *iron* and to *copper* alloys.

In the absence of any techniques for its more volumi-nous production, *steel* was never abundant in preindustrial societies, and hence it was reserved for limited special uses (swords, helmets, plowshares). But the growing supply of *iron* had enormous economic and social effects on the premodern world by greatly improving transporta-tion (horseshoes), building (axes, saws, hammers, nails), kitchen work (cookware, grates) — and warfare (guns, iron cannonballs, better firearms).

Gunpowder

Before the invention of *gunpowder,* the options to kill at long distance were few and of rather limited reach. Skilled archers using simple and compound bows, or more pow-erful crossbows (known since antiquity in both China and in the Mediterranean world), could kill men or large ani-mals at distances of more than two hundred meters. This was also the maximum reach of massive winch-powered

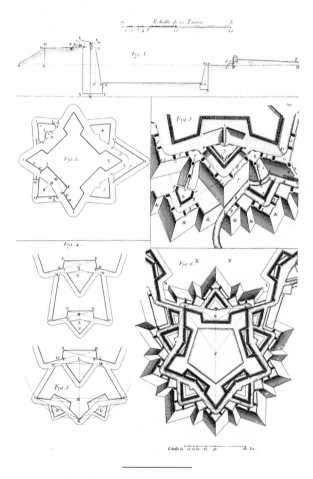

Casting of field guns became one of the first mass-production industries of the early modern world.

Huge star-shaped fortresses with massive embankments and wide ditches—whose most prolific builder was Sebastien Vauban— were designed during the seventeenth and eighteenth centuries to absorb the impacts of iron cannonballs.

After 1500 gunned sailships emerged as powerful
carriers of European technical supremacy.

catapults. Consequently, *hunters* had to come fairly close to their prey, and most combat casualties were inflicted hand to hand, by both foot soldiers and mounted warriors armed with daggers, axes, swords, and lances.

The only way to cause damage remotely resembling the effects of universally feared and revered powers of lightning was to use incendiary materials concocted with sulfur, quicklime, petroleum, and asphalt and attached to arrowheads or hurled from catapults. Such bombardments could cause a great deal of destruction in besieged settlements or garrisons, but only the formulation of *gunpowder* combined a great propulsive force with unprecedented explosive and inflammatory power.

Origins of this profound innovation were in the experience of the ancient Chinese alchemists with gunpowder's three ingredients. They knew basic properties of saltpeter (potassium nitrate, KNO_3), sulfur, and *charcoal* for centuries before an unknown Taoist monk brushed the first formula for a rather ineffective mixture of the three ingredients sometime around A.D. 850. Well-tested formulas for different *gunpowders* appeared two hundred years later, but even these mixtures were not truly explo-

sive inasmuch as they contained too little (around 50 percent) KNO_3 (highly explosive mixtures contain about 75 percent KNO_3, 15 percent *charcoal*, and 10 percent sulfur).

In contrast to *wood* or *fossil fuel* combustion, which draws oxygen from the surrounding air, ignited KNO_3 provides its own oxygen. This rapid internal supply generates instantaneously a volume of gas roughly three thousand times larger than the volume of the ignited gunpowder. The resulting force is large enough to propel even loosely packed small projectiles over considerable distances. This was proved for the first time with Chinese bamboo fire-lances of the tenth century. Once better confined and directed, the force of exploding gunpowder begun propelling increasingly heavier projectiles at longer distances. Manufacture of such guns begun in China just before the year 1200, and true guns were cast in Europe only a few decades later. More powerful and more accurate guns of the fifteenth century also used more destructive iron cannonballs.

This combination changed both the land and maritime warfare. The ability to destroy fortified stone structures erased the defensive value of castles and walled towns. Once such guns were mounted on more maneuverable *sailships*, Europe gained a highly effective means of projecting its commercial, technical, and cultural power around the world during the centuries of great colonial expansion. By the late 1600s, a century after English long-range guns defeated the Spanish Armada, great men-of-war carried up to a hundred guns. Such concentrations of explosive power were surpassed only after the middle of the nineteenth century with the formulation of nitrocellulose-based powders and, above all, with the patenting of Alfred Nobel's dynamite in 1867.

Sailships

Few energy convertors of the preindustrial era had such a profound effect on history as *sailships*. Sails—fabric aerofoils maximizing lift force and minimizing drag—cap-

A typical Elizabethan galley.

Ariel, a participant in the world's most famous 25,000-km long clipper race from Fuzhou to London in 1866.

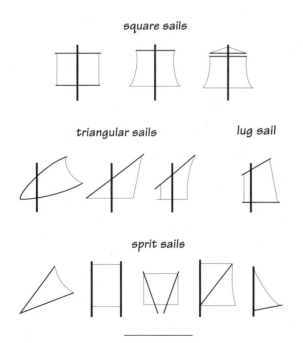

square sails

triangular sails   **lug sail**

sprit sails

Different types of sails.

tured the kinetic energy of *winds*, shaped hulls reduced their resistance, and rudders kept them on course. For millennia *sailships* provided the only means of long-distance waterborne transportation, and during their apogee their heavily armed versions became vital instruments of global empire building while their sleekest types became admirably fleeting messengers of growing international trade. By that time *sailships* could do what was impossible for their ancient counterparts: they could sail into the wind.

The basic physics involved in sailing is straightforward, but converting the kinetic energy of *winds* into forward movement when they do not blow in the direction of travel was a matter of gradual gains. The earliest ships were invariably square-sailed. Naturally, they would go speedily only with wind directly astern (180°) — Roman ships powered by northwesterlies could sail from Messina in southern Italy to Alexandria in just six days — but they could not move forward when the wind was more than thrity degrees off the desired course (150°). The best medieval square-rigged ships could sail slowly with the wind on their beam (90°). The only way to overcome this limitation was to take the best possible angle and keep changing the course: because square-rigged ships had to make a complete downwind turn, it took Roman vessels as many as seventy days to travel from Egypt to Italy, more than ten times the duration of their outbound run!

Only asymmetrical sails aligned with the ship's long axis and capable of swiveling around their masts to catch the *winds* overcame that limitation. The late medieval combination of square sails and triangular mizzen sails could handle wind at sixty-two degrees, and fore-and-aft rigs could come as close as forty-five degrees to the wind. For comparison, modern yachts can sail close to the aero-dynamic limit of thirty degrees. All ships with fore-and-aft sails could turn their bows into the wind and then catch it on the opposite side of their sails, replacing time-consuming turns by a zig-zag progress.

Europeans did not invent the essential ingredients needed to make *sailships* more efficient and more powerful — they got triangular sails from the Indian Ocean and stern-post rudders, the magnetic compass, and *gunpowder* from China — but they combined them and developed them to an unmatched degree. After centuries of stagnation, European ship design became highly innovative during the late medieval and early modern era, and a quick progress followed rapidly.

The first ship to circumnavigate the world (between 1519 and 1522) displaced a mere 85 tonnes, less than a common cargo vessel used by Romans more than a millennium earlier. But by the end of the sixteenth century naval vessels, often equipped with scores of guns, averaged more than 500 tonnes, and two centuries later British merchantmen displaced typically about 1200 tonnes. Top speeds nearly quadrupled between 1500 and 1850: in 1853 the Boston-built but British-crewed clipper *Lightning* covered 803 kilometers in one day, averaging 9.3 meters per second! Without gunned *sailships* Europe could not have accomplished its rise to global domination, and its growing fleets of merchant ships provided the foundation for intensifying world trade.

5

FOSSIL-FUELED CIVILIZATION

The single word capturing best the essence of modern high-energy societies is *growth:* growth of energy output leading to growth of cities, populations, crop yields, industries, economies, affluence, travel, information, armaments, war casualties, and environmental pollution. Given the rapidity of this growth, long-term perspectives, especially when properly interpreted, are simply stunning.

By the 1850s *wood, charcoal, and straw* were still dominant fuels everywhere except in a few European countries, and the total annual per capita combustion of all fuels was less than five hundred kilograms of *wood* equivalent. By the mid-1990s the global per capita output of *fossil fuels* and primary *electricity* prorated to about 1.5 tonne of oil equivalent a year. In gross energy terms this is a nearly eightfold rise, but because prevailing energy conversion efficiencies in 1850 averaged around just 15 percent, whereas by 1995 they reached about 40 percent, the average per capita consumption of useful energy was about twenty times higher, an unprecedented change after centuries of stagnation or marginal growth.

Moreover, a fundamental change within this change has been going on since 1900, when less than one percent of all *fossil fuels* were converted to *electricity:* By 1945 the share rose to 10 and by 1990 to 25 percent. The rapid growth of hydrogeneration and the even more rapid, but now arrested, growth of nuclear capacities further added to the availability of the most convenient of all energies as its average annual global per capita consumption more than quintupled between 1950 and 1995, from less than 400 kWh to nearly 2300 kWh.

More than any other change, these enormous energy flows transformed the twentieth century world. By transforming *traditional agriculture* they have brought huge increases in food supply. Since 1900 the world's cultivated area increased only by about a third, but with more than a fourfold increase of average yields the total crop harvest rose almost sixfold. This gain has been due largely to more than eightyfold increase of energy inputs to crop cultivation.

These energy subsidies have been used to build field machinery, power its operations, synthesize farm chemicals (above all energy-intensive *nitrogen* fertilizers) and support development of new crop varieties. Their result has been not only a surfeit of food (too rich in *lipids*) in affluent countries, but also an adequate per capita supply of basic *nutrients* in all but a few war-ravaged poor nations: remaining malnutrition is overwhelmingly the matter of access to food, rather than one of its availability.

By providing energy flows of high power density, *fossil fuels* and *electricity* made it possible to embark on a large-scale industrialization creating a predominantly urban civilization with unprecedented levels of economic growth reflected in better health, greater social opportunities, higher disposable incomes, expanded *transportation* and an overwhelming flow of *information.*

On the personal level, *electricity* has been essential in easing the lives of the traditionally disadvantaged half of the humanity as it did away with tiresome domestic *labor* and offered the possibility of female emancipation. On the national level, a close relationship between energy use and economic growth is readily demonstrated by long-term comparisons, but the growth of total energy consumption at higher stages of economic development hides impressive declines of energy/GDP intensities. Modern economies use less energy per unit of economic product because of the declining share of energy-intensive capital inputs, higher conversion efficiencies, and rising role of services. At the same time, the intensity with which modern economies use *electricity* has been increasing.

But looking ahead brings more worries than confidence. Most importantly, only about a quarter of humanity is well off, and it supports its affluence by consuming nearly three-quarters of the world's fossil fuels and primary *electricity.* In contrast, the poorest quarter of humanity can afford to buy less than one-twentieth of all commercial primary energies. Such a gap is obviously no foundation for long-term global social and political stability. Yet closing it rapidly by greatly expanded supply may cause globally unacceptable environmental impacts, above all a rapid climatic change.

A relatively fast transition to a non-fossil fueled civilization is not a realistic possibility. Historical perspectives show that the substitution of energy sources and widespread adoption of new prime movers are always very gradual processes, unfolding across several generations.

There is obviously no shortage of alternatives—from *fission reactors* to direct conversion of *solar radiation*—but their economies are still questionable and their long-term viabilities still uncertain. Fortunately, we have enormous opportunities to improve efficiencies of energy use: This path can go a long way toward redressing the great consumption gap while protecting the global environment.

Fossil Fuels and Electricity

Although modern societies do not derive all their primary energy from *fossil fuels,* combustion of *coals* and hydrocarbons (*crude oils* and *natural gases*) provides more than four-fifths of all non-biomass energy consumption. Nearly all of the rest is supplied by primary *electricity* gen-

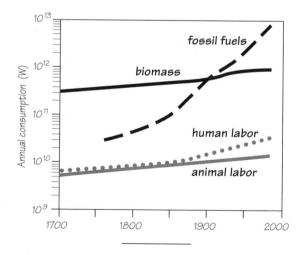

Energy transition from biomass to fossil fuels.

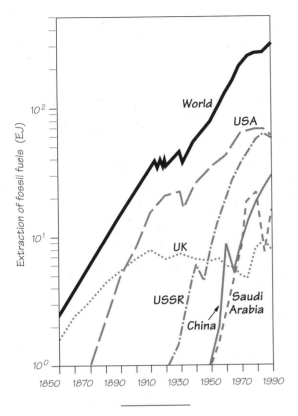

Global extraction of fossil fuels.

troduction of *steam turbines,* first for the generation of *electricity; internal combustion engines,* which took rapidly over most means of transportation; and *electric motors,* which revolutionized industrial production and became indispensable for many other productive and household tasks. With efficient *transmission, electricity* also provided superior *lights* and it made possible such key advances as the production of *aluminum* and the synthesis of *nitrogen* fertilizers. Further prime-mover revolutions came with introductions of *gas turbines,* which transformed aviation, and *rockets.* These made space flight a reality—but they also can deliver *nuclear weapons* in intercontinental ballistic missiles.

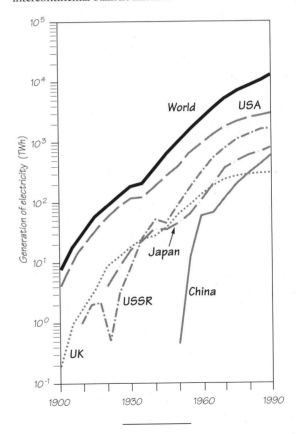

The first century of electricity generation.

erated by *water turbines* and *fission reactors.* In Western Europe the transition from *wood, charcoal, and straw* to mineral fuels as the main sources of heat, and from a heavy reliance on animate *labor* (supplemented by *waterwheels* and *windmills*) to fossil fuel-fired mechanical prime movers, was well under way by 1850 (when coke displaced *charcoal* in *blast furnaces*) and it was virtually completed before World War II. In contrast, the poor, populous countries of Asia began this shift only after 1950, and most countries in sub-Saharan Africa have barely started it.

Steam engines were the first fossil fuel-fired prime movers, but their dominance ended quickly with the in-

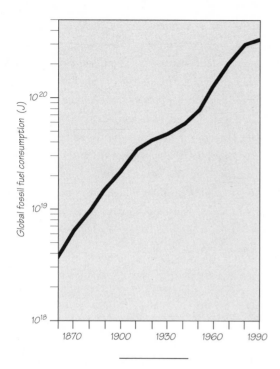

Rising per capita consumption of primary energy.

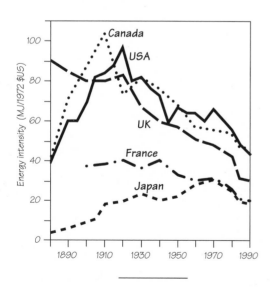

Impressive improvements in energy intensity of major economies.

Coals

Combustion of *coals* had energized the birth of *fossil-fueled civilization* and creation of modern world—and it still contributes about a third of the global primary energy consumption. These solid fuels have a common origin in the lithification of peats produced by accumulations of dead plant matter in wetlands. Differences in original vegetation and, more importantly, in magnitudes and durations of transforming temperatures and pressures, have produced a large variety of *coals.*

Anthracites and the best bituminous *coals*—ancient fuels derived primarily from the *wood* of large, scaly-barked trees laid down in immense coastal swamps more than two hundred million years ago—break up to reveal hard, jet-black facets of virtually pure carbon. These fuels contain hardly any water and very little ash, and their

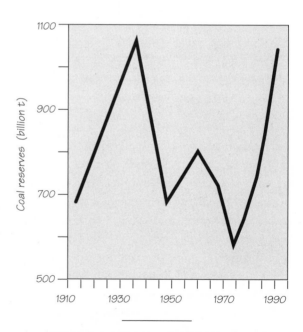

Global estimates of coal reserves have not changed substantially since the first published value in 1913.

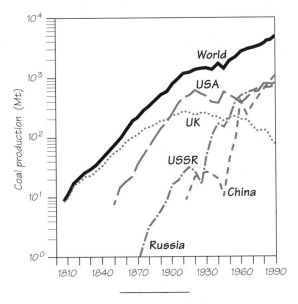

Two centuries of coal extraction.

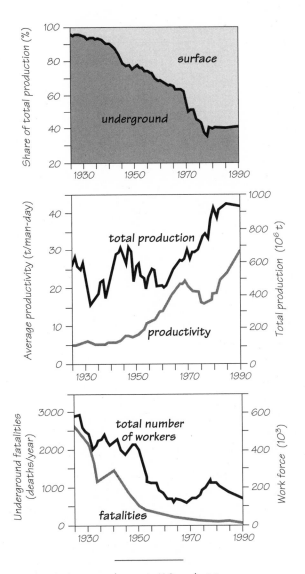

heat values are near 30 MJ/kg. In contrast, the youngest lignites are soft, often crumbly, clayish, fairly wet chunks of brownish hues. They contain large shares of ash and water, often a great deal of sulfur, and their energy density is as low as 10 MJ/kg, inferior to air-dried *wood.*

Not surprisingly, all large lignite-producing nations call them brown *coals.* As with their drawbacks, the redeeming qualities of brown *coals* spring from their relative immaturity: their deposits are often very thick, quite level and close to the surface, conditions making a large-scale extraction in huge open-cast mines very economical.

Earlier in this century the first large surface mines worked the deposits where the overburden/seam ratios were no higher than 1.5–2; now they have risen commonly to 4–5. Surface lignite mines are the sites where the world's largest earth-moving machines clear away one, two, and three hundred meters of overburden before large excavators bite into *coal* seams.

Long-term changes in U.S. coal mining.

The largest open-cast mines claim over 10 km² of land, move annually over nearly two hundred million tonnes of earth and extract ten to forty million tonnes of fuel, an energy equivalent to some nationwide deep coal-mining totals! Even larger projects would be technically feasible inasmuch as there are huge untapped deposits of brown *coal* in Asia and in Africa. Russians could develop mines with capacities up to sixty million tonnes along the Trans-Siberian railway between Kansk and Achinsk. Nearly a quarter of all recoverable global *coal* reserves is in lignites, but in energy terms this share shrinks to less than one-seventh.

Some brown *coals* still serve as feedstock for synthetic chemicals, but this use had its most momentous past during World War II, when hydrogenation of German lignites helped to prolong the war. After the loss of Romanian oilfields (the Red Army took them on August 30, 1944) the Wehrmacht's trucks and Luftwaffe's planes continued to run until April 1945 largely on synthetic gasoline.

Today most of the fuel is burned in large mine-mouth plants to generate the *electricity* that runs the computers of the German federal government in Berlin, lightens the gloom of cloudy Moscow evenings, and illuminates the interiors of Prague's splendid baroque churches. Germans, Russians, and Czechs rely heavily on lignites, producing about two-thirds of their global output.

The high sulfur content of European lignites produces large emissions of sulfur dioxide. When converted to sulfates and transported by prevailing *winds* for hundreds of kilometers, these emissions are a leading source of acid deposition affecting sensitive ecosystems from Scandinavia to Switzerland.

The eastern part of North America has a similar problem with emissions from large-scale combustion of bituminous *coals*. As in Europe, North America's *coal* market is heavily dominated by *electricity* generation, with production of coke, the leading fuel in *iron* smelting, far behind. The United States still leads the world in large-scale mining of high-quality bituminous *coal,* with most of the output coming from highly productive, surface operations. Typically, they recover around nine-tenths of all fuel

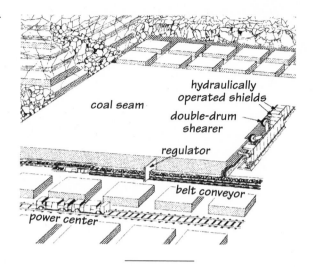

Long-wall mining, the most efficient method of underground extraction.

present in the seam, and daily productivities can surpass thirty tonnes per worker.

In contrast, underground mining leaves up to half of the *coal* behind, and its daily productivities may be just a few tonnes per miner. Productivity is even lower in China, now the largest producer of raw *coal.* Roughly half of its output is a low-quality fuel from small rural pits, and only about one-fifth of total extraction is sorted and cleaned before combustion.

There are no near-term limits on the global availability of *coals.* Measured recoverable reserves, in seams exceeding thirty centimeters and no deeper than 1,200 meters, surpass 700 billion tonnes, equivalent to about 250 years of the 1995 rate of extraction. But better quality fuels — *crude oils* and *natural gases* — have been displacing *coal* for decades, and environmental concerns, especially the worries about CO_2-induced global warming, make any large-scale *coal* resurgence unlikely.

Crude Oils

Crude oils are surprisingly uniform fuels. Their energy density fits mostly in the range of 42–44 MJ/kg, 50 percent above that of standard *coal,* and their ultimate analysis shows narrow spans for the dominant carbon (84–87 percent) and hydrogen (11–14 percent). Their diverse colors and densities — ranging from brownish, light, and volatile liquids to black, heavy, and viscous oozes — stem from different shares of paraffins, cycloparaffins, and aromatics. Thomas Gold argues that they could have originated not only from surface life but also from the metabolism of *bacteria* in the deep and hot biosphere extending several kilometers below the *Earth*'s surface.

Unfortunately, most *crude oils* are relatively heavy and sour (containing more than one percent of sulfur), and they require expensive catalytical cracking to produce higher shares of lighter refined fuel fractions, and desulfurization to prevent high SO$_2$ emissions. Light and sweet (low-sulfur) crudes that have high yields of gasolines, jet fuels, and kerosene even with simple thermal distillation

Eleven of the world's fifteen largest oil fields are in, or adjacent to, the Persian Gulf.

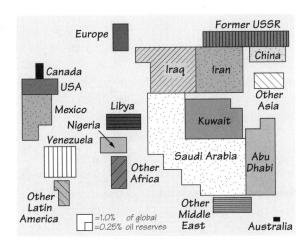

Shares of global crude oil reserves in 1995 are shown by proportionate areas of national or continental territories.

are produced in quantity only by Algeria and Nigeria, and they command premium prices.

In the Middle East *crude oils* have been known for millennia from natural seepages and pools, but they were used only rarely as fuels (heating of Constantinople's *thermae* during the late Roman Empire is perhaps the best example), and more frequently as protective coatings. The initial impetus for large-scale *crude oil* extraction was the replacement of expensive whale oil in lamps by kerosene distilled for the first time in 1853 by Abraham Gesner.

Pioneering American enterprise, which began with Colonel Edwin Drake's Pennsylvania well in 1859, was followed rapidly by many major discoveries. By 1900 two other American states (Texas and California), Romania (the Ploesti fields), Russia (the huge Baku fields on the Caspian Sea), and Dutch Indonesia (Sumatra) led the ranks of *crude oil* producers. During the first two decades of the twentieth century they were joined by Mexico, Iran, Trinidad, and Venezuela (1914). By that time electric *lights* made kerosene an archaic illuminant and the main demand for crude oil products was coming from the rapid adoption of *internal combustion engines* powering *cars, ships, planes,* and agricultural machinery.

Global *crude oil* extraction grew tenfold between 1920 and 1960, when oil surpassed *coal* to become the world's leading source of *fossil fuels.* This ascent was supported by post-1945 discoveries of giant oilfields in the Middle East, Siberia, Nigeria, Indonesia, Mexico, and Alaska, and it led to the development of extensive networks of *pipelines* and to the emergence of global oil trade using ever larger *tankers.* Between 1945 and the early 1970s *crude oil* combustion fueled the world's longest

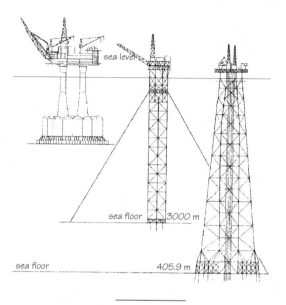

Offshore drilling rigs are among the largest, as well as the most massive, artifacts of fossil fueled civilization.

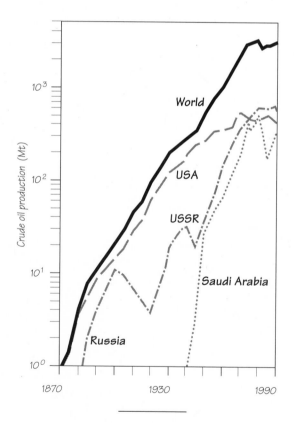

Global crude oil extraction.

sustained period of economic expansion, which propelled the United States, Canada, most of Europe, and Japan to affluence and which launched many poor countries, most notably in East Asia, on a path of swift modernization. Crude oils products gained large new markets in numerous chemical syntheses, in *electricity* generation, in diesel-powered *trains* and in aviation based on progressively more powerful *gas turbines.*

As the frequency of giant field discoveries declined while global demand kept rising, the Organization of Petroleum Exporting Countries took advantage of the extraordinary concentration of the world's *oil* reserves in the Persian Gulf basin and began to make spectacular profits after it quintupled its *oil* price in 1973 and then it tripled them in 1979 and 1980. OPEC's fabulous earnings, amounting to the largest and fastest international transfer of money in history, and its enormous global influence, were short-lived. After 1985 world *oil* prices resumed

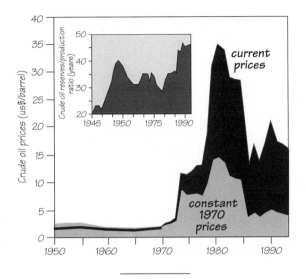

Long-term trends in international crude oil prices. *Inset:* Fluctuations of global crude oil production/reserve ratio since 1945.

their historic decline, and by the mid-1990s the global reserve/production ratio rose to almost fifty years, the highest level since 1945.

But the underlying resource realities have not changed. About thirty thousand hydrocarbon fields have been discovered since the drilling of Drake's Oil Creek well in nearly three hundred continental and offshore basins. Most fields have only minor deposits, and almost four-fifths of the world's *crude oil* reserves are in just five basins: the Persian Gulf area, Western Siberia, the Gulf of Mexico, Russia's Volga-Ural region, and Venezuela's Maracaibo. Even more impressively, five Persian Gulf countries — Iran, Iraq, Kuwait, Saudi Arabia, and the United Arab Emirates — have two-thirds of all global *crude oil* reserves. Saudi Arabia alone controls one-fourth, and its largest field has one-seventh of the Earth's known deposits.

Global estimates of ultimately recoverable *crude oil* indicate that the production could be sustained at the mid-

1990s level for at least 80, and perhaps for up to 120, years. This span could be extended by extracting more costly oils locked in shales and sands. In any case, we will try to use oils as long as possible: their high energy density, easy transportability and convenient storage make them the quintessential energy source for modern *fossil-fueled civilization.*

Natural Gases

Extraction of *natural gases* has a curiously early but isolated ancient beginning and a delayed but steadily rising modern importance. *Gases* were burned in China's landlocked Sichuan province since the beginning of the Han dynasty (200 B.C.) to evaporate brines. Their extraction

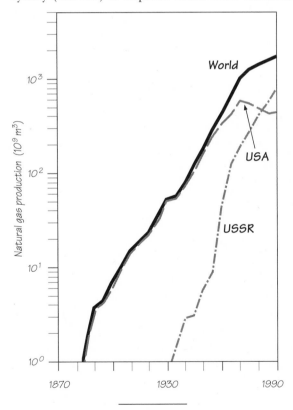

The first century of natural gas extraction.

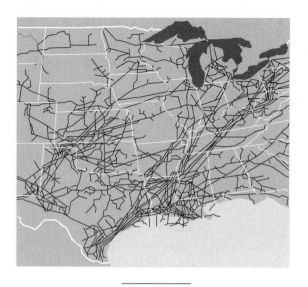

America's dense network of natural gas pipelines.

No matter if the gases are associated with *crude oils* or if they are extracted unaccompanied, their composition is always dominated by methane (typically 75–95 percent), and the mixtures contain declining shares of higher alkanes (ethane, propane, butane) and frequently traces of CO_2, H_2S, N, He, and water vapor. Pure methane has energy density of 35.5 MJ/m³, and the highest values for *natural gases* are about 25 percent above that rate. The cleanliness of *natural gases*—absence of fly ash and the relative ease with which impurities can be removed from contaminated alkanes—makes them the best fuel for heating and cooking. They have also become indispensable in a large number of chemical syntheses; most notably, methane is both the principal fuel and feedstock in producing synthetic *nitrogen* fertilizers.

The medium estimate of ultimately recoverable global *natural gas* resources is only about 15 percent lower than the total for *crude oils,* and the maxima are virtually identical. Because of still substantially lower rates of annual *gas* extraction, the worldwide reserve/production ratio was more than sixty years at the mid-1990s

was made possible by the invention of percussion drilling: An *iron* bit was fastened to a bamboo cable suspended from a bamboo derrick and was raised by men jumping on an attached lever. The deepest boreholes produced by this simple technique reached only ten meters during the Han dynasty, surpassed a hundred meters a millennium later, and culminated in the one-kilometer well in the early nineteenth century. *Natural gas* was led in bamboo pipelines and burned under large cast *iron* pans filled with brine. Some lighting and cooking uses are also documented in classic Chinese literature.

After 1900 expanding *crude oil* production was accompanied by substantial volumes of associated *natural gas,* but most of this excellent fuel was simply flared: its modern importance came only after World War II with widespread installation of high-pressure *pipelines* and powerful compressors that made the long distance, high-volume distribution of *natural gases* economical. Global extraction rose more than tenfold between 1950 and 1970, and then it doubled by 1990.

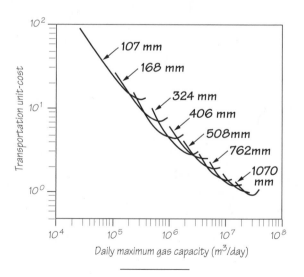

Gas transportation cost as a function of pipeline capacity.

rates, and the maximum resource estimate would give the *fossil-fueled civilization* more than two hundred years of *natural gas* use.

Distribution of *natural gas* reserves is no less uneven than that of liquid hydrocarbons. Russia has about one-third of global reserves, the Persian Gulf countries about two-fifths. Urengoi and Medvezhye, Russia's largest fields in Western Siberia, discovered in the late 1960s and producing since the late 1970s, contains about one-tenth of the world's proved *natural gas* reserves. Their enormous flow is led through the world's longest *pipelines* all the way to Central and Western Europe as far as southern Germany and northern Italy. *Natural gases* from rich North African, Southeast Asian, and Alaskan fields are liquefied and transported by special *tankers* to Western Europe, the east coast of the United States, and Japan.

Steam Engines

The steam engine was the first effective, economic, and durable convertor of the inanimate chemical energy in fuels to kinetic energy. Its invention opened up the way for transforming the world's rich deposits of *coals* into reciprocating and rotary motion: the engine became the first new prime mover since the early medieval diffusion of *windmills.*

The machine's practical development started in 1690 when Denis Papin experimented with a toylike benchtopmodel. Shortly afterward Thomas Savery built a small (about 750 W) steam-driven pump operating without a piston, and by 1712 Thomas Newcomen introduced a much better, and five times more powerful, engine for water pumping in English mines. Its principal drawback was its very low efficiency. Condensing the steam on the underside of the piston left only about half a percent of *coal's* energy to be converted to reciprocating motion.

John Smeaton's persistent efforts roughly doubled this poor efficiency, and Newcomen's pumping engines

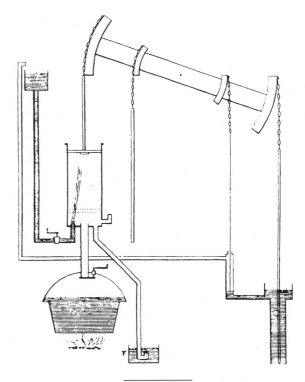

Savery's simple atmospheric engine of the early eighteenth century.

were installed in a growing number of English mines after 1750. The beginnings of James Watt's fundamental innovation go back to his work at Glasgow University in 1763–64 when he had to repair a working model of Newcomen's engine. He knew little about *steam engines* and confessed that he approached his task "as a mere technician"—but by 1765 he had a clear idea of what had to be changed.

His addition of a separate condenser was patented on January 5, 1769, and the patent was subsequently extended by the Steam Engine Act of 1775 to the end of the century. Watt's other critical innovations consisted of insulating the cylinder by a steam jacket and maintaining a vacuum in the steam condenser with an air pump. Later improvements included a double-acting engine with a piston driving also on the downstroke, and the famous centrifugal governor maintaining constant speed with varying engine loads.

The combination of Watt's engineering skills and Matthew Boulton's capital resulted in the construction of some five hundred reciprocal and rotary **steam engines** by the year 1800. Although the mean power of their machines (about 20 kW) remained rather small, it was almost three times as large as that of average contemporary **windmills,** and more than five times that of typical **waterwheels.** The engine's locational flexibility was no less important than its growing capacities. Watt's largest units, rated at just over one hundred kilowatts, merely matched

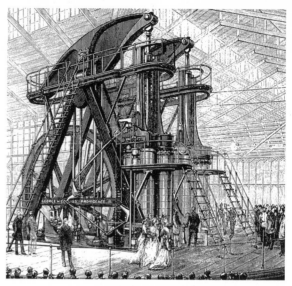

Huge (1.1 MW) Corliss engine at the Philadelphia Centennial Exhibition of 1876.

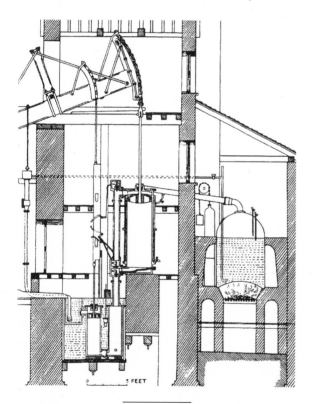

Watt's engine with a separate condenser patented in 1775.

the largest contemporary **waterwheels**—but they could be sited in any place that could be supplied by **coal** and by the relatively small volumes of water needed for steam generation.

Advances in the engine's mobile applications had to wait until the expiration of Watt's patent. Although he was keen to improve his stationary low-pressure engines, Watt tried to block any development of high-pressure machines for locomotives. In 1784 he rejected the proposal by William Murdock, the foreman in charge of installing his engines, to form a new partnership for the manufacturing of locomotives based on Murdock's design, and two years later he actually tried to prevent Murdock's patent application. Watt's distrust of high-pressure steam and his sweeping patent rights thus delayed practical advances in mobile **steam engines** until after 1800.

Subsequent innovations resulted in machines that were not only much more powerful but also considerably

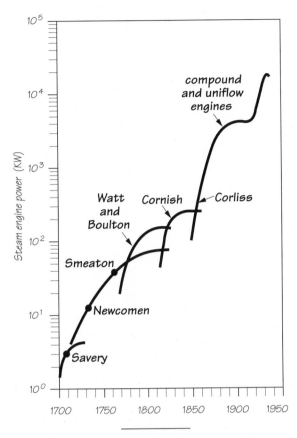

In more than two hundred years the largest steam engine capacities increased by four orders of magnitude.

By the end of the nineteenth century, the power of the largest machines (cross-compound Corliss engines) approached five megawatts, the highest operating pressures were about a hundred times higher than during Watt's time, the best efficiencies surpassed 20 percent, and the weight/power ratio of the lightest engines declined to less than a hundred grams per watt, about one-fifth of its average 1800 value.

In spite of these enormous advances *steam engines* could not meet new requirements. Locomotives remained fairly inefficient (typically just between 4 and 8 percent), even the lightest engines were too heavy to power fast and easily maneuverable road vehicles, and even the best

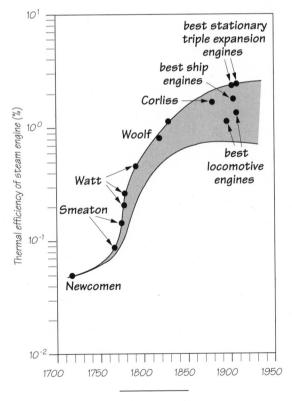

The best steam engine efficiencies rose about fiftyfold between 1700 and 1930.

more efficient and more versatile. The *steam engine* became the quintessential machine of the nineteenth century industrial expansion, the dominant prime mover in all kinds of stationary applications ranging from powering mine pumps and winches and *blast furnace* bellows to turning belt drives in spinning factories and in countless other manufactures. After 1830 steam engines also rapidly revolutionized both land and water transportation. *Trains* and steam-powered *ships* created not only new national and regional, but also, for the first time in history, reliable transcontinental and global networks of transportation.

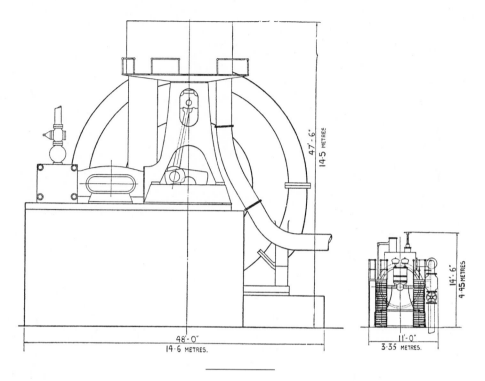

Parsons's 1906 comparison of sizes of equally powerful
(3.5 MW) steam engine and steam turbine.

machines were too large to serve as prime movers in rapidly expanding coal-fired electric stations. Steam engines thus could not compete either with much more powerful *steam turbines* in *electricity* generation, or with much lighter *internal combustion engines* in road transport, and soon also in *airplanes.*

Steam Turbines

From refrigerating vaccines and food to landing *airplanes,* from synthesizing ammonia to disseminating *information,* our societies are utterly dependent on *electricity*—but the ubiquitous prime mover used to generate about four-fifths of its global supply is perhaps the most obscure of all the critical nineteenth-century mechanical inventions that have transformed twentieth-

century civilization. Few people have ever seen a *steam turbine,* and most of those aware of its existence can offer just a rudimentary black-box description.

Its invention could not have been more perfectly timed. *Steam engines* rotated the first *electricity*-generating dynamos, but were an unsatisfactory prime mover for this purpose. Inherently massive, they could not be scaled up without reaching enormous sizes and consuming huge amounts of materials; relatively slow, they could not deliver high rotations desirable for electric generators; in spite of many improvements they were still relatively inefficient; they would be rather uneconomical in large-scale generation.

Steam turbines offered superior alternatives in all three respects. Their much smaller size for identical power

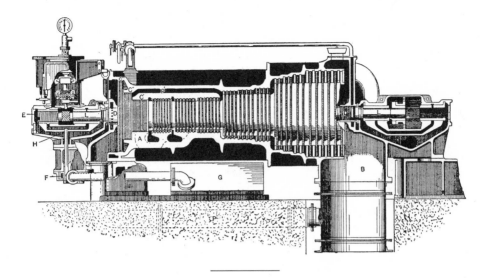

Section of Parsons's first commercial steam turbine,
a 75 kW machine built in 1887.

rating was featured by Charles A. Parsons, their inventor, in his company's early advertisements: while the weight/power ratio was about 250–300 g/W for the best *steam engines,* even early steam turbines rated well below 50 g/W. Parsons's first commercial turbine operated at 4800 rpm, while reciprocating engines usually delivered less than 100 rpm. And higher operating pressures and temperatures translated into higher efficiencies and into potential power ratings orders of magnitude above those for practical *steam engines.*

Not surprisingly, the development and commercial diffusion of steam turbines was exceptionally fast. Charles Parsons demonstrated the first reaction turbine design in 1884, and his company installed a 75 kW turbine at a public power station in Newcastle in 1888 and the first stationary condensing turbine of 100 kW in 1891. In 1897 a 44-tonne *Turbinia* equipped with a 1.5 MW turbine proved the superiority of the machine in marine propulsion as it outraced every warship in a naval review, and in 1900 the two sets for Elberfeld station in Germany be-

came the first turbogenerators rated at 1 MW. Subsequent steady increase in highest ratings was interrupted between the early 1930s and the late 1940s, but the exponential rise resumed in the early 1950s, and by 1965 both the largest sets and the modal size of newly ordered units rose by an order of magnitude.

This rapid growth—from maxima of 150–200 MW in 1955 to 1000–1300 MW by 1970—was made possible by a combination of economic incentives, technical improvements, and system considerations. Pronounced economies of scale in terms of both capital and operation costs favored larger turbogenerator ratings. Further savings were realized with installations of several identical units (boiler-turbogenerator combinations) at one site.

Better materials could meet the required design specifications. The earliest machines were made of cast *iron;* C-Mo *steel* could accommodate temperatures around 480°C by the late 1930s; addition of chromium raised throttle temperatures to 530°C and austenitic *steel* brought it to just above 560°C. These advances can be also easily traced

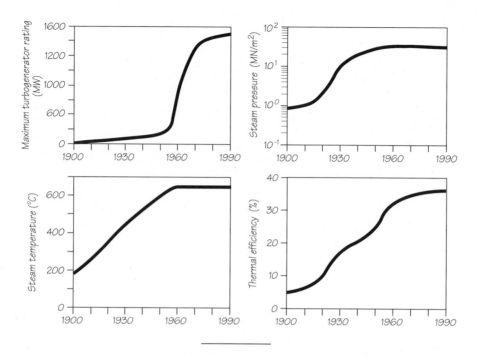

Historical development of U.S. fossil-fueled
turbogenerators, 1900–1990.

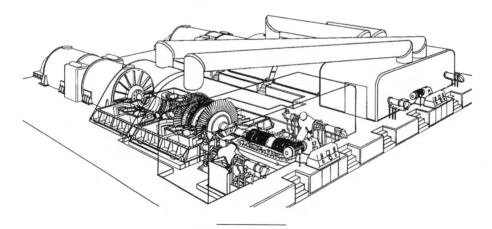

Cutout view of 1.3 GW turboset, one of the
world's largest, built by Brown Boveri in 1971.

by the length of rotor blades, growing from less than twenty centimeters around 1910 to fifty-five centimeters by 1947 and to one meter by the early 1960s. And because the largest generating unit should not surpass 5 percent of a network's total installed capacity, expanding *transmission* links made it possible to opt for larger unit sizes without imperiling the reliability of a system's generation.

After 1970 the much lowered demand for new generating capacity in virtually all rich countries shifted the emphasis from higher ratings to higher efficiency and reliability. The largest installed turbogenerators have reached a plateau at about 1.5 GW, and the mean size of new units has actually declined to below 500 MW. Both of the world's largest turbogenerators, the 1450 MW unit at Ignalina in Lithuania and the 1457 MW unit at Chooz in France, use steam from *fission reactors;* the largest coal-fired units have capacities just short of 1.4 GW.

Top performances have been achieved by a series of thermodynamic refinements. The most efficient arrangements involve supercritical double-reheat units. They operate at temperatures above the critical point of water (374°C), and they return the steam from very high- and high-pressure segments of the turbine to, respectively, high- and intermediate-pressure rotors after reheating it in superheater. The maximum theoretical efficiency of this turbine cycle is just above 50 percent, and as the best supercritical turbines are close to 90 percent efficient in transforming the kinetic energy of expanding steam into rotation, the best possible thermal efficiencies are around 45 percent.

Water Turbines

The first radical and commercially successful improvement of hydraulic prime movers since the early medieval diffusion of vertical *waterwheels* came only in 1833, when Benoit Fourneyron won the competition for a new water-

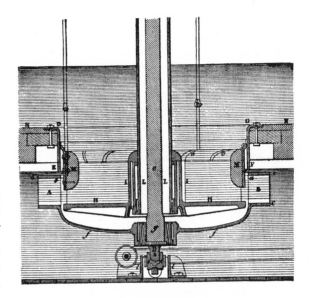

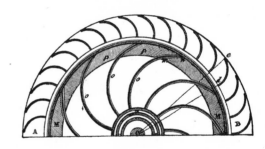

Section and plan of Benoit Fourneyron's 1832 turbine for Freisans ironworks.

powered prime mover sponsored by the Société d'Encouragement pour l'Industrie Nationale. Conceptually derived from earlier, and much less efficient, designs of horizontal *waterwheels* with curved blades, his reaction *turbine* was driven by radial outward flow. By 1837 two of his machines for Saint Blaisien spinning mill worked under unprecedented heads of 108 and 114 meters, developing almost 45 kW, enough to run thirty thousand spindles and eight hundred looms.

Fourneyron's *water turbines* were fairly efficient (above 80 percent), but they performed well only under particular flow and pressure conditions. Their success spurred a great deal of interest in hydraulic design, and they were soon eclipsed by a variety of better machines. The very first one of these innovations — an inward-flow turbine of James B. Francis patented in 1849 — has been by far the most important hydraulic prime mover. The spiral casing of decreasing diameter and adjustable guide vanes lead water onto completely submerged curved runner blades; after transferring its kinetic energy to the rotor, the water flows away through the central outlet.

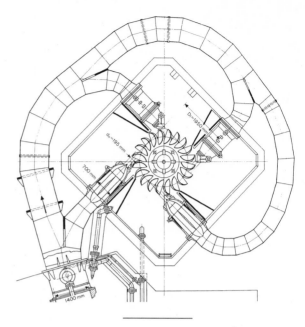

Horizontal section through a vertical-shaft, four-nozzle Pelton turbine by Escher Wyss (rated at 37 MW, head 415 m).

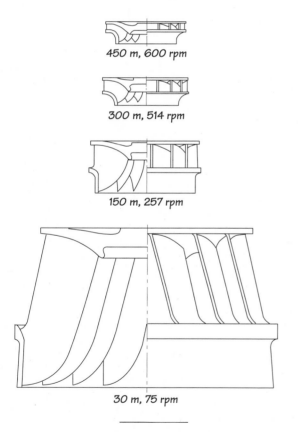

450 m, 600 rpm

300 m, 514 rpm

150 m, 257 rpm

30 m, 75 rpm

Relative increase of Francis turbine runners with decreasing generating head.

Francis *turbines* are suited for a wide range of generating heads: those around one hundred meters are most common, but modern large machines are installed at sites with as little as thirty meters and as much as about three hundred meters. With higher heads they have been replaced by impulse wheels. Runners of impulse *turbines,* patented by L. B. Pelton in 1889, have closely spaced smooth bilobal buckets. Each bucket cuts off a portion of water jet with a minimum of shock loss, turns it through a nearly 180° angle, and evacuates the spent water with the lowest possible loss of potential energy. Operating arrangements can range from wheels on horizontal shafts propelled with a single jet (typical for the highest heads) to vertical shafts with up to six jets on a single runner.

Various fixed-blade propeller *turbines* are particularly suitable for hydrostations with large natural storages guaranteeing reliable year-round flow, but providing only

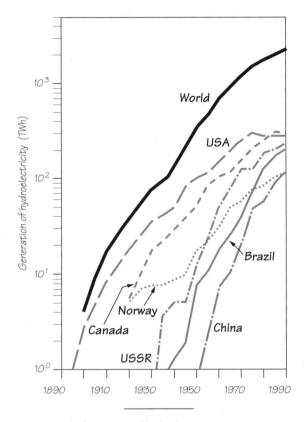

The first century of hydroelectric generation.

tions, and by 1895 the first 3.72 MW unit of a ten-unit Niagara plant began generating. Post–World War I development was boosted first by the Soviet quest for rapid industrialization (summarized by Lenin's slogan, "Communism equals the Soviet power plus electrification"), and then by extensive U.S. federal involvement in building dams and electrifying the countryside as a means of economic recovery during Roosevelt's New Deal.

Until the 1980s the USSR and the United States remained the builders of the world's largest (above 2 GW) hydrostations, equipped with increasingly more powerful generating units (above 500 MW). The record sizes were eventually reached at America's Grand Coulee Dam with 700-MW Francis *turbines* and total plant capacity of 6.8 GW. This record was surpassed by Brazil and Paraguay's Itaipu Dam (12.6 GW); in turn, this will be topped by China's Sanxia Dam (17 GW) sometimes around 2010.

Water turbines now account for nearly one-fourth of global generating capacity, and they produce one-fifth of the world's *electricity.* Among rich countries, hydrogeneration is especially important in Canada, the United States, and Russia, but it is relatively much more important in many relatively small producers throughout the poor world, where it supplies the bulk of available *electricity:* shares are in excess of 90 percent in many African countries, and 80–90 percent in South America.

While European, North American, and non-Siberian Russian resources have been largely exploited, enormous generating potential in Asia, Latin America, and Africa guarantees further substantial growth of hydrogeneration. These resources include not only sites suitable for multi-GW stations, but tens of thousands of local opportunities for tapping hydropower with small *water turbines* in stations with capacities ranging from just a few tens of kW to a few MW. Since the 1960s China has led such efforts, and its small-scale hydro capacity reached about 19 GW by 1990.

relatively small differences in elevation (commonly less than 20 m). The Kaplan *turbine,* introduced in 1920, is a special kind of a propeller machine: the angle of its runner blades is adjustable to fit best the desired load or the operating head.

First *water turbines* powered directly a variety of industrial processes, most notably textile mills, but their importance rose rapidly once they were coupled with *electricity* generators (in the United States for the first time in 1882, the same year as thermal generation), and once the advances in *transmission* made it possible to harness water resources distant from major load centers. By the early 1890s Switzerland had some two hundred hydrosta-

Fission Reactors

In 1971 Glenn Seaborg, chairman of the U.S. Atomic Energy Commission, opened the Fourth International Conference on the Peaceful Uses of Atomic Energy by offering a number of predictions for the year 2000. Nuclear energy would be a phenomenal success bringing unimagined benefits for the greater part of humanity; it would provide half (or about 750 GW) of the U.S. *electricity* generating capacity, and similar shares in other affluent countries; nuclear *tankers* and explosives would be widely used, nuclear-propelled spaceships would ferry men to Mars, and energy from nuclear fusion would be well on its way to commercialization. A generation later none of this has come to pass, and the retreat of nuclear generation has been almost certainly the costliest technical miscalculation of the twentieth century.

Soon after Seaborg's Geneva speech the future of nuclear energy seemed even more promising as OPEC's rapid *crude oil* price increases and embargoes seemed to made fission an even more desirable option: It could end OPEC's extortions and do so not only as a lower price — but also with much less environmental impact than coal-fired generation. Worldwide, the industry had some three hundred *reactors* under construction or in various stages of planning and its expansive future seemed assured.

In reality, it began a costly, and still far from finished, retreat caused by a combination of causes. Its development was far too rushed, in terms of both installed capacities and, especially in the United States, the number of plants under construction. The first commercial *fission reactor* began generating *electricity* in 1956, just fourteen years after the first sustained nuclear chain reaction on December 2, 1942, and during the 1960s aggressive orders pushed the average capacity of a new nuclear power plant over 1 GW. As the U.S. utilities opted for many individualized designs, rather than for a standard reactor, and as they began to build concurrently dozens of new plants, con-

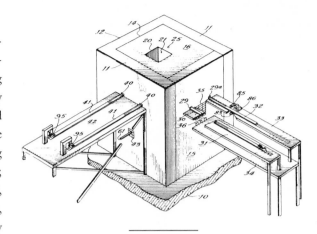

Drawing of the first nuclear reactor, Chicago Pile No. 1, from a patent application by Enrico Fermi and Leo Szilard (filed in 1944, issued in 1955).

struction costs escalated, and they were pushed up further by changing requirements of new safety regulations.

Public acceptance of nuclear generation, never better than lukewarm, declined with growing realization of potential risks of routine reactor operation and of intractable difficulties of what is in civilizational terms a virtually eternal (10^4 years) storage of spent radioactive fuel; naturally, it plunged after the 1979 Three Mile Island accident. Collapse of high *crude oil* prices in 1985 and much lower demand for *electricity* made the economics of nuclear fission even more dubious.

Effects on the industry were truly devastating. U.S. utilities chose not to complete stations which have already cost them billions of dollars; Germany stopped any further construction, Swedes voted to close their plants in the future, Austrians did not start up the one they actually finished. In the early 1980s only France, USSR, and Japan were proceeding with extensive plants, but completions of previously ordered plants kept pushing up the share of nuclear generation in the world's total output of *electricity* from less than 2 percent in 1970 to 8 percent in 1980 and 15 percent in 1985, with France reaching about 70 percent and Sweden and Taiwan surpassing 50 percent.

The Chernobyl disaster in 1986, the universally unresolved problem of long-term storage, new studies showing very high costs of reactor decommissioning, and the demise of the Soviet empire had largely closed the chapter on the first generation of *fission reactors.* Most of them are pressurized water reactors (PWR) whose coolant circulates through the core in a closed loop and transfers its heat to a *steam turbine,* whereas in boiling water reactor (BWR) steam is generated by passing water directly between the elements of nuclear fuel. In both of these designs ordinary water acts both as a coolant and as a moderator slowing the neutrons and thus improving the efficiency of conversion. In Canadian pressurized heavy-water reactors D_2O is a moderator, and British gas-cooled reactors use graphite as moderator and He or CO_2 as coolant. None of these designs has a bright future. The industry's only realistic hope remains in demonstrating an

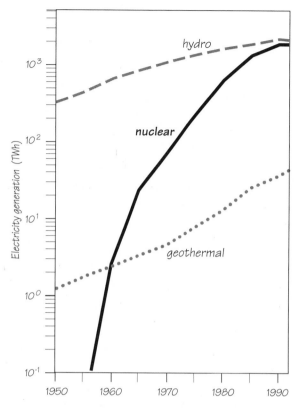

Global nuclear electricity generation compared to other sources of primary electricity.

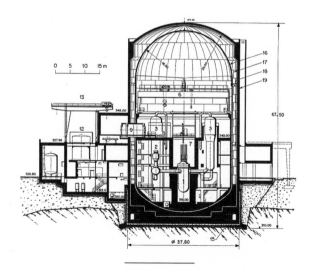

Section of a containment building of a 350 MW PWR reactor in Beznau, Switzerland.

inherently safe reactor design coupled with a convincing solution for long-term waste disposal.

And what of the dreams of unimagined benefits for most of the humanity? Since 1945 hundreds of billions of dollars have been invested in research and development of nuclear generation, but in the mid-1990s all operating *fission reactors* produced about eight EJ of *electricity*— while about four times as much energy was consumed, mostly in poor Asian, African, and Latin American countries, as *wood, charcoal, and straw.* Even when correcting this comparison by appropriate conversion efficiencies— more than 90 percent of *electricity* can be turned into useful energy, while biomass fuels are rarely burned with efficiencies surpassing 25 percent — the poor world's fuel-wood, crop residues, and dung still delivered about as much useful energy as the world's *fission reactors.*

Transmission

When in 1882 Thomas Edison designed the first commercial electric system centered on the Pearl Street power plant, he chose direct current (DC) at 110 volts as the safest way of transmitting *electricity.* Fires and electrocutions experienced with alternating current (AC) circuits built for arc-light systems were a clearly on his mind, and many experienced scientists and engineers, including Lord Kelvin and Werner von Siemens, agreed with him. But just four years later George Westinghouse, a newcomer to the business of *electricity,* began marketing systems using AC voltages above one kilovolt and planning connections with up to ten kilovolts.

What came to be known as "the battle of the systems" was an emotional, even nasty, and very public dispute

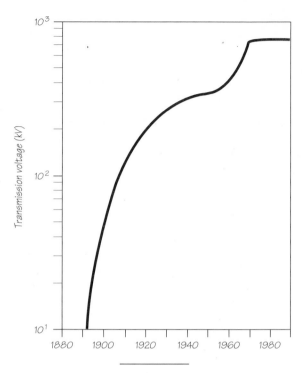

Growth of the highest AC transmission voltage
in the United States, 1886–1990.

dominated by Edison on one side and by George Westinghouse, Nikola Tesla, the inventor of first practical *electric motors,* and Sebastian Ferranti on the other. Edison was unyielding and blunt, forecasting ("just as certain as death") that Westinghouse would kill a customer within six months of installing his AC lines. But in this case the great innovator was on the losing side of a critical contest. Before the 1880s ended two inventions opened up much wider prospects for transmitting AC *electricity:* The transformer made it much more economical, and Tesla's polyphase *electric motors* introduced a huge new demand. Edison's stubbornness was not self-destructive. Once the AC had clearly made it on the market, Edison's usual innovative acumen led him to cut his losses, and his company stopped the manufacture of DC components.

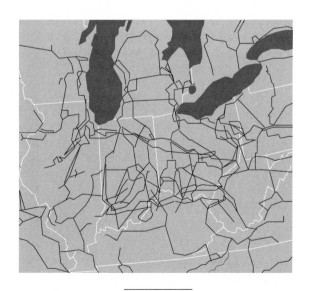

High-voltage power grids in America's heartland.

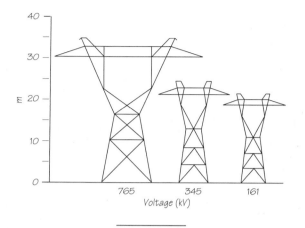

Typical high-voltage transmission towers.

Installation of first reliable and inexpensive transformers by William Stanley in 1886 was by far the most important advance in the *transmission* of *electricity* during its first formative decade—and the device has remained one of the critical infrastructural ingredients of modern civilization. These conceptually simple, yet flexible and durable devices, convert high current and low voltage into low current and high voltage and vice versa. Electric current generated at relatively low voltages could be stepped up for more economical transmission, and then stepped down for customized distribution. Without transformers it would have been necessary to minimize *transmission* distances, and the resulting decentralization of power generation would have made it impossible to harness impressive economies of scale associated with larger sizes of *steam turbogenerators, water turbines* and *fission reactors.*

First American networks started with voltages of just 4.6 and 6.9 kilovolts, but before World War I the highest long-distance ratings rapidly progressed to 23, 69, 115, and 138 kilovolts (all multiples of the final distribution voltage of 115 volts). Rapid growth of voltages resumed in the early 1950s, and within about two decades the highest standard voltages settled at 345, 500, and 765 kilovolts. Inevitably, all *transmission* components had to be changed greatly in order to keep up with higher voltages and higher power ratings.

Wires changed first from solid to stranded *copper;* the same sequence was repeated with *aluminum.* Economical long-span wires were made possible by stranded *aluminum* conductors reinforced with cores of stranded *steel.* Supports evolved from simple wooden poles with wooden crossarms to tall *steel* and *aluminum* towers, initially designed just for function, later also for better appearance.

Since 1900 the power of largest transformers has increased about five hundred times, their voltage rose about fifteen times, their weight/power ratio fell by an order of magnitude, and their efficiency rose to more than 99 percent.

And DC *transmission,* requiring only two insulated conductors and hence costing only about two-thirds the price of a three-conductor AC link transmitting the same

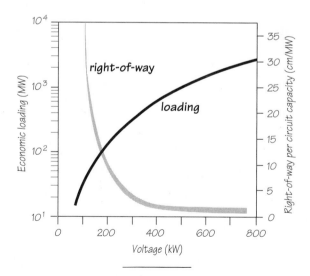

Decrease in rights-of-way and increase of power capacity with higher voltage.

amount of power, returned to provide some of the world's most important links, either over relatively short distances by undersea cables (Denmark–Sweden, New Zealand's two islands, England–France) or between large, remote hydrogenerating stations and major urban load centers. The latter links began in 1970 with the West Coast Pacific Intertie (carrying 1.44 GW at 400 kV over 1330 km) and with Manitoba's (1972–77) Nelson River-Winnipeg link (1.62 GW at ±450 kV over 890 km). In the late 1980s the record passed to the 800 km ±600 kV link between Itaipu, the world's largest hydrostation, and Sao Paulo. Still, most of the substantial international *electricity* trade (particularly important in Europe) is carried by AC lines.

In spite of many technical advances, *transmission* remains the most expensive way of long-distance energy distribution. Even the second most expensive means, moving *natural gas* by *pipeline,* costs only about a seventh as much. And getting new rights-of-way has been increasingly difficult at any cost as traditional land use and esthetic objections have been overshadowed by concerns about health effects of low-level electromagnetic fields, although the most feared risk, a higher incidence of some childhood cancers, has been identified only in the most tenuous statistical terms.

Electric Motors

By far the largest share of the world's total prime mover capacity is installed in *internal combustion engines,* above all in *cars;* the world's most powerful individual prime movers are *steam turbines* (for sustained conversion) and *rocket engines* (for enormous but brief thrust). But none of these prime movers would have risen to its respective prominence without *electric motors.* In many ways their introduction was even more revolutionary than the introduction of *steam engines* because of the profound change of power distribution and control.

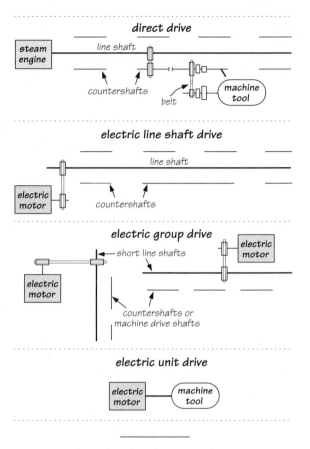

Evolution of machine drive in manufacturing.

Coal-fired steam generation introduced a radically new kind of prime mover—but the distribution of steam-generated mechanical power for processing and machining did not differ from the pattern established by *waterwheels:* gears rotated long *iron* or *steel* line shafts passing along factory ceilings, and pulleys and leather belts transmitted this motion to parallel countershafts belted to individual machines, often also to different floors through holes in the ceiling. Any mishap—prime-mover breakdown, damaged gearing, cracked shafts, slipped belts—stopped the whole assembly. And the whole arrangement

Small electric motors with the bearings located in the end bracket are ubiquitous industrial prime movers.

was inherently inefficient, as there was no way to power only selected machines.

First *electric motors* in factories were installed to drive shorter shafts powering smaller groups of machines, but the unit drive began to dominate after 1900. Consequences of this change revolutionized modern manufacturing. Gone were large gearing and the friction losses and frequent system outings inherent in centralized power distribution. Unit power supply allowed flexible plant design, and by removing the overhead shaft-and-belt clutter with its noise and accident risks it freed the ceilings for better *lights* and ventilation, a change resulting in greater workplace safety and higher labor productivity. Additional productivity gains and enormous improvements in the consistency and quality of finished products came from precise machine control possible with individual *electric motors.*

The motor's principle was appreciated for decades before the first commercial generation of *electricity:* conversion of electric current into rotary motion is simply the reverse of electromagnetic induction demonstrated by Far-

aday in 1824. And after Z. T. Gramme built his first reliable ring-shaped armature dynamos in 1870 he could easily demonstrate that a pair of them could run interchangeably as a generator or a motor — but only with direct current. Edison's company began marketing small DC motors soon after it began generating *electricity* in 1882, and by 1886 Frank Sprague designed much larger (up to 11 kW) DC motors — but with the victory of alternating current *transmission* most of the market rapidly shifted to Nikola Tesla's three-phase AC motors, patented in 1888 and offered for the first time by Westinghouse in 1892.

AC motors were not only simpler, smaller and lighter than DC motors delivering the same output, they also did not spark, needed hardly any maintenance and were much cheaper to make. During the 1890s they began replacing *steam engines* in countless factory operations, and in the United States this prime-mover revolution ran its course by 1930. During the first three decade of the century capacities of *electric motors* in American factories rose nearly sixtyfold to just over four-fifths of the total installed power. This share has changed little since that time.

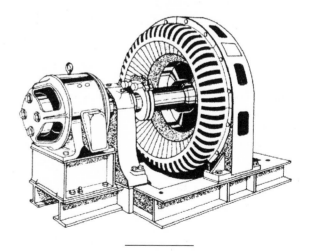

A large synchronous electric motor.

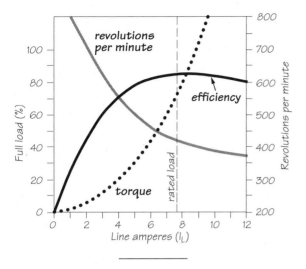

Load characteristics of a DC motor.

The induction motor of Tesla's invention remains by far the most common AC device for converting *electricity* into rotary motion. Two or more coils of its stator produce a magnetic field inducing a current in its rotor; the turning force on the rotor continues until the rotor's motion matches the speed of the magnetic field. When a nearly perfect constant speed is required synchronous motors are used, while series motors can operate from either AC or DC supply, and because of their high starting torque they are the best choice for electric locomotives. Modern electronic speed control techniques have made it possible to regulate all types of motors with a great precision.

Concurrently with their conquest of factory floors, *electric motors* were making inroads into new markets ranging from mechanization of household chores (washing machines and vacuum cleaners) to railway transportation (the first electric *trains* were introduced during the 1880s). Their versatility and reliability was demonstrated by applications involving tiny units operating dental drills and large motors opening massive lock gates of the Panama Canal.

This breadth of applications has only increased with time, embracing frivolous uses (opening car sunroofs) and life-saving applications (moving blood in heart-lung machines), ubiquitous tasks (circulating chlorofluorocarbons in refrigerators), and extraordinary missions (powering the world's largest bucket excavators in surface *coal* mines).

Lights

Some stories can be reread many times without any loss of excitement. Thomas Edison's account of the first working light bulb is in that special category. After hundreds of hours of failed experiments with different materials cotton

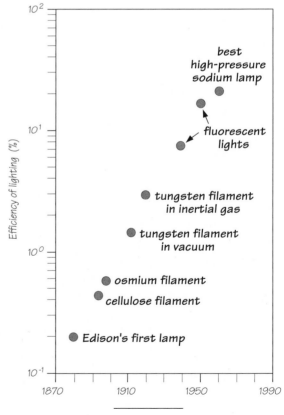

Increased efficiency of lighting.

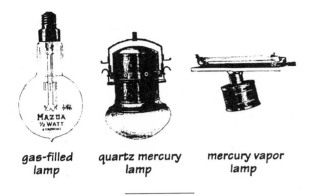

gas-filled
lamp   quartz mercury
lamp   mercury vapor
lamp

Three different types of lamps from the 1920s.

sewing thread was carefully carbonized, inserted into a glass evacuated by suction pump and connected to *electricity* supply from a dynamo. The light bulb "burned like an evening star," and we have the exact beginning of a new era for civilization: October 21, 1879.

Soon the thread was replaced by oval loops of carbonized paper secured by little platinum clamps, then by carbonized bamboo fiber, squirted cellulose, and tantalum. In 1912 came ductile tungsten filaments; since 1913 the bulbs have been filled with an inert gas (argon or krypton) in order to reduce filament evaporation, and the filament was coiled. Since 1934 the filament has been produced as a coiled coil to reduce heat loss from convection. Each step was accompanied by increased luminous intensity, higher efficacy (the quotient of the total luminous flux emitted by the rated power, measured in lumens per W) and longer expected life for mass-produced bulbs.

Carbonized fibers gave off no more than 2.5–3 lm/W, tungsten filaments generated 10 lm/W in a vacuum and 20 lm/W in gas-filled bulbs, and the average rated lifetime rose from less than three hundred to a thousand hours. Half a century after Edison's successful experiment the technique of producing incandescent lamps was mature and little improvement lay ahead.

Today's standard hundred-watt bulb, frosted inside for better light diffusion, is good for 750 hours and it puts out initially 17.5 lm/W. The most efficient incandescent light, a 10 kW source for film and TV studios, lasts only twenty-five hours and puts out 33.6 lm/W, but it is still highly inefficient. Taking 1.47 mW/lm as the standard mechanical equivalent of light means that this powerful lamp converts no more than about five percent of *electricity* into light. The efficiency of common hundred-watt bulbs is less than 3 percent.

Major efficiency gains came only with discharge and fluorescent lights. Discharge lamps are filled with either inert gas (most commonly neon) or metal vapor (mercury or sodium) producing blue or yellow light suitable for outdoor illumination. In fluorescent lamps special compounds coating the inside of glass absorb UV rays generated by excitation of low-pressure mercury vapor and reradiate it as illumination approximating daylight. Efficiencies of both kinds of lamps are relatively high. Standard forty-watt fluorescent tube converts nearly 12 percent of *electricity* into visible light, and it lasts at least twenty-five times longer (more than twenty thousand hours) than a hundred-watt incandescent light. High-intensity

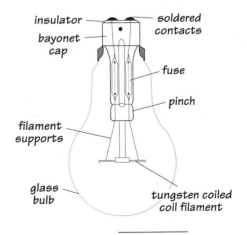

Incandescent light bulb.

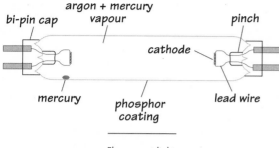

argon + mercury
vapour

bi-pin cap

pinch

cathode

mercury

phosphor
coating

lead wire

Fluorescent light.

discharge metal halide (putting out up to 110 lm/W) and high-pressure sodium lamps (up to 140 lm/W) convert around 20 percent of *electricity* into visible radiation.

An incandescent bulb can be seen as an inefficient holdover from a simpler era—but its cheapness and ease of installation guarantees its continuing popularity even in rich countries. And there are hundreds of million people in Asia, Africa, and Latin America who have yet to see its warm light changing their lives as it changed ours.

Internal Combustion Engines

Papin's late seventeenth-century experiments included not only trials of toy *steam engines* but he also continued Huygens's tests of small devices powered by exploding *gunpowder.* Steam power took off almost immediately with Newcomen's machines, but the first practical *internal combustion engine,* fueled with a more suitable explosive mixture of gas and air, was designed only more than a century and half later by Etienne Lenoir. Other, mostly oil-fueled, prototypes were built soon afterward by Hock, Brayton, Capitaine, and Hornby during the 1860s and 1870s, but the best design turned out to be a four-stroke *coal* gas-fueled horizontal engine by Nikolaus Otto, patented in 1878.

Possibilities of widespread commercial application opened up only with Gottlieb Daimler and Wilhelm Myabach's improvement as gasoline-fueled, light, high-speed,

single-cylinder vertical engine patented in 1885. Its advantages soon conquered the early enthusiasm for electric *cars,* and the gasoline engine became a signature prime mover of the twentieth century. More than a billion of them have been installed, mostly in *cars* and also in motorcycles, trucks, *airplanes,* boats and in a growing range of stationary and mobile gadgets ranging from small *electricity* generators to lawn mowers.

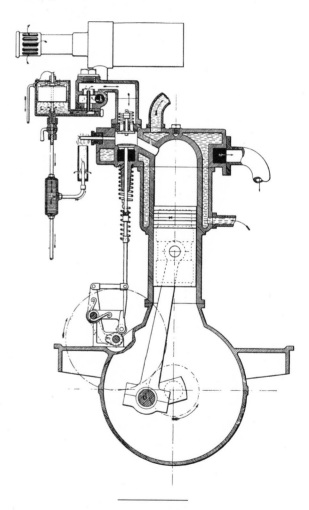

Section of an 1899 Daimler gasoline engine with float-feed carburetor and hot-tube ignition.

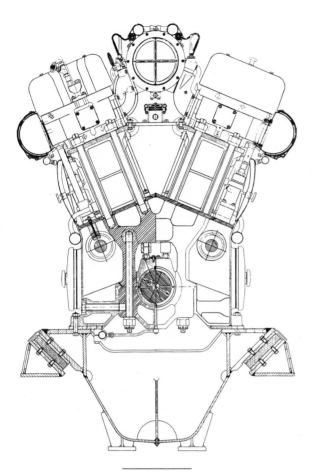

Section of a large German-made locomotive diesel engine.

and notable declines of engine weight/power ratios or, inversely, higher specific power output.

Typical compression ratios have more than doubled, to 8–9 in the 1990s, and even subcompacts have engines more than three to four times as powerful as Ford's celebrated model T, rated at fifteen kilowatts. Most of the weight/power decline happened during the first generation of engine's development. Otto's 1880 engine weighed nearly 270 g/W, Daimler and Maybach's radical redesign cut it to 40 g/W a decade later, and some of the first mass-produced engines released after 1900 weighed just 5 g/W.

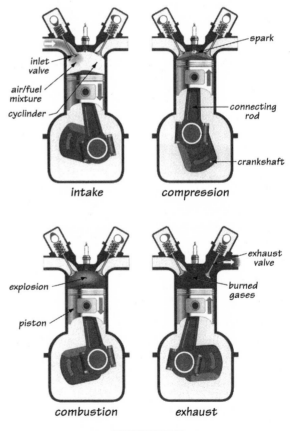

Four-stroke-cycle engine.

There have been many impressive gains in the engine's efficiency, reliability, and average power during the first century of its use — but Daimler would not be at loss when inspecting the most advanced designs of the late twentieth century because the development of Otto cycle engines has been, on the whole, rather conservative.

The three principal trends of improvement have included the rise of compression ratios, substantial increase of average engine power in its most common applications,

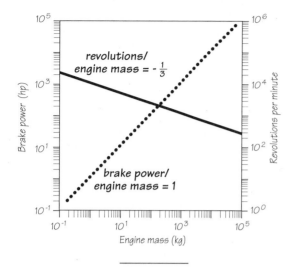

revolutions/
engine mass = $-\frac{1}{3}$

brake power/
engine mass = 1

Brake power (hp)

Revolutions per minute

Engine mass (kg)

In spite of the different numbers of cylinders and types of fuel and ignition, there are clear allometric relations between power, rpm, and the mass of internal combustion engines.

Nine decades later typical truck engines weigh as little 3 g/W, and those in passenger *cars* rate just around 1 g/W.

Aircraft engines became almost that light just after a decade of flying. The Wright brothers' home-made engine was rather heavy at 9.1 g/W, but V-12 Liberty engines mass-produced in the United States during World War I weighed merely 1.34 g/W. By the late 1950s, when powerful piston engines in commercial aircraft began to be displaced by *gas turbines,* their weight was down to less than 0.6 g/W.

By that time gasoline powered engines were also losing some of their earth- and water-bound market to a different *internal combustion engine,* to Diesel engine patented in 1892. Because of its high compression ratio (15–24) the injected fuel ignites spontaneously but the engine is necessarily heavier. But its higher weight makes little, or no, difference in many applications, and in a number of other uses this drawback is more than compensated

for by the engine's inherently superior efficiency and by its use of cheaper liquid fuels.

The best diesels have efficiencies of up to 42 percent, and performances above 30 percent are routinely achievable with any well-maintained machine. In contrast, although the ideal efficiency of Otto cycle is 60 percent, the best practical performance of gasoline engines is only around 32 percent. The advantage of using heavier liquid fuels in diesel engines is obvious: Most *crude oils* are made up of relatively heavy fraction and can produce higher yields of gasoline only with expensive catalytic cracking.

Diesels made their first major inroads in marine propulsion, both in *ships* and submarines, during World War I, and their dominance in this market (except for nuclear propulsion in the largest submarines) was complete by 1960. On land they began to be used first in heavy earth-moving and farming machines. As they became lighter (5–10 g/W), they displaced *steam engines* powering locomotives, and even lighter machines, weighing as little as 3–4 g/W, conquered most of the world's rapidly growing truck and bus transport.

Although there is no shortage of diesel-powered *cars* (whose engines weigh between 2–5 g/W of shaft power), gasoline engines have remained dominant in that market. Only in *airplanes* could diesels not compete: The best-ever diesel aero engine, Germany's World War II 745 kW Jumo 207, weighed 0.87 g/W, still too heavy to compete with sparking machines. A new field opened up for diesels as *electricity* generators, either as standbys in central stations or in remote locations where *transmission* would be too expensive.

Gas Turbines

In these distinct types of *internal combustion engines,* compression of gas precedes combustion, and the expansion of hot gas emerging from the chamber drives a

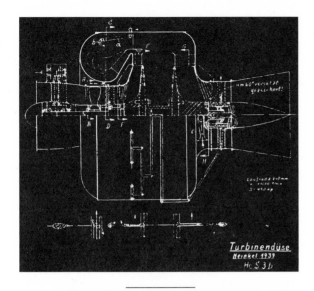

Pabst von Ohain's original 1939 cross-section
of gas turbine designed for a Heinkel fighter.

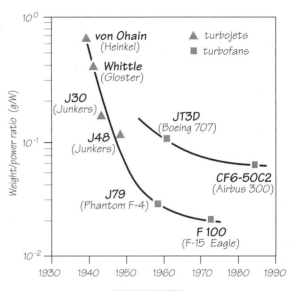

Declining weight/power ratios of gas turbines used
in commercial and military aircraft.

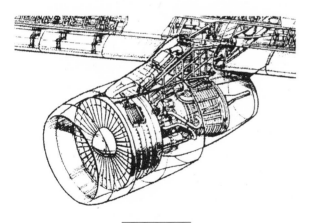

Pratt & Whitney's PW4000 turbofan jet engine
mounted on a wing of Boeing 747.

turbine. If a turbine is just large enough to turn the compressor, most of the emerging hot gas can generate a powerful forward thrust in a turbojet engine. Alternatively, after powering a compressor hot exhaust gases can be used to drive propeller shafts (on ships or planes) or *electricity* generators. The first practical application of *gas turbines* came only in the late 1930s, when independent work by Frank Whittle in England and Hans Pabst von Ohain in Germany led to the testing of prototypes powering new military *airplanes.*

Jet aircraft saw only limited service near the end of World War II, but the low weight/power ratio of *gas turbines* and their high ratio of thrust per frontal area made them the best choice for airborne prime movers and led to their rapid postwar advances. *Gas turbines* first displaced reciprocating engines in fighters and bombers, and in the late 1950s they revolutionized long-distance commercial aviation.

A succession of new jet *airplanes* called for more powerful — and, after the sharp oil price rise in 1973, more efficient — engines. The most powerful jet engine ever built, the GE4/JSP, was to power the supersonic Boeing B2707–300SST. At Mach 2.7 it would have delivered

248.32 MW — but the abolition of the U.S. supersonic transport program in 1972 ended the prototype's development. With the exception of a few supersonic Concordes, powered by a Rolls Royce/SNECMA turbojet, the most common class of large engines are those with turbofans powering the still growing fleet of jumbo *airplanes.*

In these engines some of the air compressed by an additional set of fans bypasses the engine to mix with hot combustion gases: The increased volume of exhaust gases generates higher thrust. These engines are capable of more than 200 kN of thrust, and they can develop up to 60 MW. The weight/power ratio of these large commercial turbofans is just 0.06–0.07 g/W, and their thrust/weight ratio has surpassed 6. In contrast, the best fighter engines weigh just below 0.02 g/W, and have a thrust/weight ratio around 8.5.

Gas turbines have found many useful niches beyond spectacular applications in aviation. Relatively small turbines (mostly between 1 and 15 MW) drive centrifugal compressors moving *natural gas* in *pipelines,* maintain pressure in oil wells, compress gas for *blast furnaces* and for chemical syntheses (including production of *nitrogen* fertilizers), generate *electricity* by using waste gases in refineries and *steel* mills, and power heavy *trains* and *ships.*

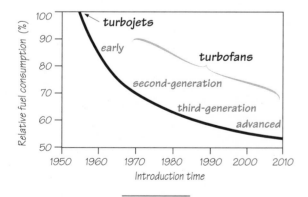

Relative fuel consumption of turbojets and turbofans.

Much larger turbines (commonly 20–80 MW, with maxima near 150 MW) power *electricity* generators used not only for standby and peaking service, but also increasingly for the base load. Their flexibility and low capital cost are their main advantages, but typical efficiencies of the open-cycle generation are below 30 percent, lower than those of *steam turbines.* This performance can be boosted to about 45 percent with combined generation cycles. In this technique hot (almost 600°C) turbine exhaust gases generate steam, which drives a steam turbine. Using aeroderivate stationary *gas turbines* is another promising option: although more expensive, they are more efficient and easier to repair. And micro-sized *gas turbines* (25–250 kW) are now used to supply power to isolated communities or to cover emergency and peak demand.

Blast Furnaces

Ours is still very much the Iron Age — and *blast furnaces* remain at the core of the vast enterprise producing the metal. Global output of pig *iron* is still about a dozen times as large as the combined total of five other leading metals. *Steel,* made from pig *iron* by reducing its high carbon content by alloying and by physical treatment, remains the mainstay of modern infrastructures, from rails to *steam turbines* generating *electricity,* from pressure vessels of nuclear *fission reactors* to reinforcing bars in concrete.

Large modern *blast furnaces* are among the most remarkable artifacts of industrial civilization. They are slightly conical shafts whose height is just over 30 m and whose internal volume is mostly in the range 2,000–4,000 m³. At their bottom is the circular hearth where the liquid *iron* and slag collect between tappings, and in whose upper part are tuyeres, openings admitting pressurized hot air blast. The bosh is a rather short, truncated, outward-sloping cone within which are the furnace's highest temperatures; the shaft is the longest and slightly narrowing

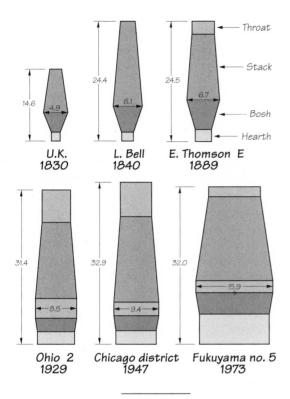

Evolution of modern blast furnaces.

Mass and energy flows needed to operate large *blast furnaces* are prodigious. A furnace producing ten thousand tonnes of metal a day will need more than 4.5 megatonnes of hot air blast a year, and it has to be supplied with about 1.6 tonnes of pelletized ore, 400 kilograms of coke, 100 kilograms of injected *coal* (or 60 kg of fuel oil), and 200 kilograms of limestone for every tonne of iron. This adds up to annual mass of 8 megatonnes of raw materials and close to 60 PJ of energy, an equivalent of nearly 2.5 megatonnes of good steam *coal*. The world's largest *blast furnaces* consume *fossil fuels* and *electricity* with power densities of about 12 MW per square meter of their hearths. Such a huge continuous *heat disposal* is of the same order of magnitude as the *solar radiation*

part where the countercurrent movement of downward moving ore, coke, and limestone and upward moving hot, CO-rich gases reduce oxides into *iron;* the top is surmounted by an apparatus for charging raw materials into the furnace, and by the pipes for gathering and conducting waste gases.

Blast furnaces can operate under pressures of up to about 250 kPa, and they may produce hot metal continuously for up to ten years before their refractory brick interior and their carbon hearth are relined. The maximum daily output of the largest furnaces is over ten thousands tonnes of hot metal, which is then converted in basic oxygen furnaces into *steel;* traditionally it was first cast into sand beds, forming the "pig."

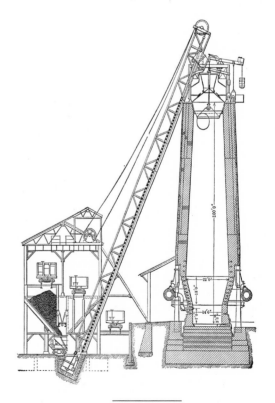

Section through an early twentieth-century furnace.

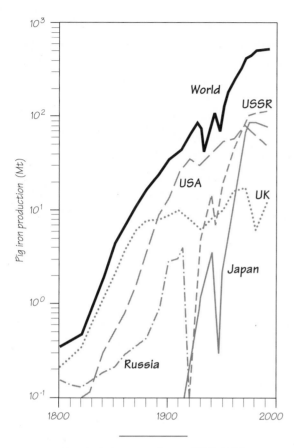

Growth of pig iron production, 1800–1990.

coke and limestone may be over by the end of the century, after almost exactly 250 years of growth.

But a decade later it became clear that neither the rapidly expanding scrap-based production of *steel* in new minimills, nor the promising but still much less important direct reduction of *iron* (by 1995 it supplied well below 5 percent of the world's pig *iron*) will be able to eliminate *blast furnace* smelting for at least another generation. The resilience of this old technique owes to continuous innovations resulting in impressive economies of scale. Between 1750 and 1976 the internal volume of the largest furnace grew from twenty to just over five thousand cubic meters, the hearth area was enlarged by two orders of magnitude (to almost 200 m²), and the highest daily production rates rose from around three to just over twelve thousand tonnes.

No less impressive than these exponential increases has been the concurrent rise in energy efficiency. The

leaving the star's photosphere (64 MW/m² if isotropically distributed), and on the *Earth* it is equaled only by power densities inside the boilers of our largest thermal *electricity*-generating stations.

Americans displaced the British as principal ironmaking innovators before the end of the nineteenth century, and they were in turn displaced by the Japanese after 1960. Since that time every one of the successive sixteen furnaces holding the world record for internal volume was built in Japan. Shortly after the world's largest blast furnace of 5,070 cubic meters was blown-in at Kyushu Oita Works in 1976, it appeared that the smelting of *iron* with

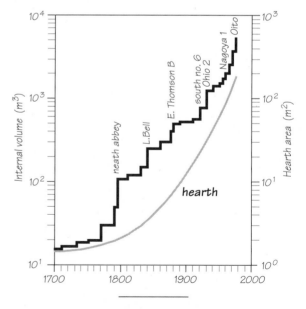

Long-term trends of the largest blast furnace capacity and hearth area.

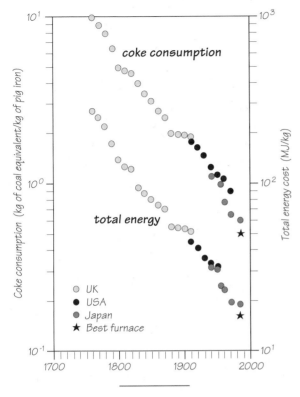

Declining use of energy in pig iron smelting.

slightly over one-tenth of the worldwide flow of **fossil fuels** and primary **electricity**. These energy savings have not meant only cheaper production but also above all greatly reduced environmental burdens caused by **coal** mining and coking.

Aluminum

The only metal to challenge **iron** in both large-scale structural applications and in numerous small-scale manufactures became available in commercial quantities only during the last decade of the nineteenth century. **Aluminum** was isolated by H. C. Oersted in 1824, but the metal remained an oddity (sometimes used for a novelty jewelry) for more than sixty years. Even the independent discoveries of its large-scale production in 1886—by C. M. Hall in the United States and P. L. T. Heroult in France—

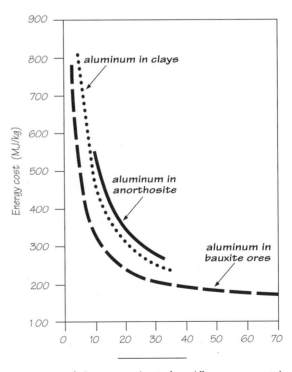

Energy cost of aluminum production from different raw materials.

average coke consumption per kg of **iron** has been declining exponentially from about ten kilograms of **coal** equivalent in the mid-eighteenth century English furnaces to an average of only 0.6 kilograms by 1990. Of course, in modern furnaces part of the coke is substituted by other liquid fuels or by pulverized **coals** injected through tuyeres, but the total energy cost of pig **iron** shows an identical rate of decline, from nearly 300 MJ/kg of hot metal 250 years ago to less than 20 MJ/kg.

This huge decline in energy intensity of pig **iron** smelting means that its production now consumes less than 5 percent of the global use of primary commercial energy; if the energy intensity had remained at the level reached by 1900, ironmaking would be now consuming

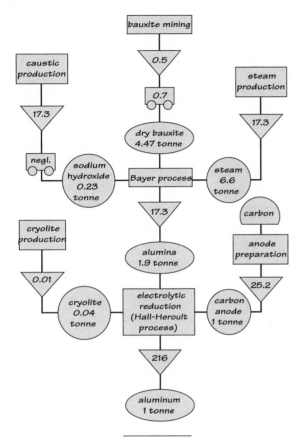

Material inputs and energy costs of
aluminum production from bauxite.

their wooden age into a new era of sleek *aluminum* bodies during the late 1920s. This demand rose sharply with the manufacture of tens of thousands of fighters and bombers during World War II, and it was sustained not only by the postwar growth of aviation, but more generally by the rise of Western prosperity.

Aluminum alloys (mainly with Cu, Fe Mn, Mg, Si, and Zn) have made substantial inroads beyond aviation wherever the design required a combination of lightness and strength. Excellent corrosion resistance is another advantage. Transportation industries became the metal's largest users, increasing its share in *cars* and *trains,* and turning its alloys into such diverse final products as recreational boats and space vehicles. This substitution process is still very much in progress: a typical car will have about twice as much *aluminum,* mostly in its engine, transmission, and small parts in the year 2000 as it did in 1990.

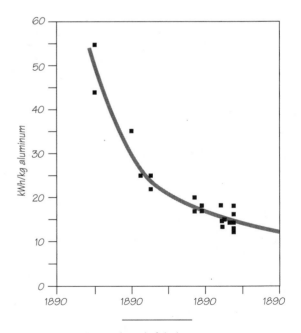

Historical trend of declining energy
cost of the Hall-Heroult process.

would have made little difference for its supply if inexpensive *electricity* had not become available from first large hydrostations.

Hall-Heroult electrolysis of alumina—produced from bauxite, *aluminum*'s most common ore—dissolved in molten cryolite was highly energy-intensive: its initial version required more than 50 MWh of *electricity* for a tonne of metal, more than six times as much energy as needed to smelt a tonne of *iron.* Consequently, *aluminum* smelting advanced only slowly, and the first large market was created only when *airplanes* emerged from

Other major *aluminum* uses are in construction (from skyscraper claddings to window frames), *transmission* (its high conductivity makes it an excellent choice for high-voltage cables), food packaging (from now ubiquitous soft drink and beer cans, which have become about 25 percent lighter since 1960, to various foils), and in the manufacture of a large assortment of industrial and consumer parts, utensils (as an excellent heat conductor it makes efficient cookware), gadgets, and appliances. By 1962 its worldwide production surpassed 5 million tonnes, and by 1995 it reached twenty million tonnes.

This expansion was greatly helped by steady improvements to the Hall-Heroult process. By the early 1990s they lowered its *electricity* requirement by about three-quarters compared to the 1900 rate. But the gap between the energy costs of *aluminum* (200–300 MJ/kg) and *steel* (mostly 20–50 MJ/kg) remains large. This explains why the metal is relatively the most widely recycled material: its production from scrap costs only a fraction of energy needed to extract it from bauxite. Titanium, which has been replacing *aluminum* in high-temperature applications, above all in supersonic aircraft, is at least three times as energy-intensive, but it can withstand temperatures more than three times as high.

Nitrogen

Nitrogen's biospheric presence is small but indispensable. The element is part of the DNA and RNA nucleotides storing and transferring all genetic information needed for the building of proteins, complex molecules made up of amino acids that serve both as structural components and as vital messengers, receptors, markers, defenders, and catalysts of life processes. Because we cannot synthesize essential amino acids, we have to consume them preformed in food proteins. Consequently, an adequate supply of *nitrogen* for crops — providing directly as plant foods and indirectly as meat, eggs, and dairy products most of en-

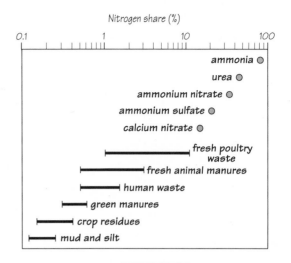

Nitrogen content of common organic and inorganic fertilizers.

ergy for our *metabolism* — is a key prerequisite of heterotrophic existence and growth.

Concerns about the adequacy of available *nitrogen* supply arose with the realization of the element's critical role in food production and of its biospheric scarcity and mobility. By 1900 there was a feeling of urgency among the agronomists and chemists aware of the looming *nitrogen* barrier to higher crop productivity. Intensive recycling of organic matter could produce high yields, but there was not enough recyclable waste to supply the *nitrogen* needed for large-scale increases of average yields. Both the inorganic *nitrogen* from Chilean nitrates ($NaNO_3$) and organic guano from the Chincha Islands could provide only a temporary reprieve. Energy needs for the synthesis of cyanamide ($CaCN_2$), introduced in Germany in 1898, and for the production of *nitrogen* oxides in an electric spark, began in Norway in 1903, were too high to become widely used.

The fundamental breakthrough came only with the invention of ammonia synthesis. Fritz Haber devised a promising experimental process to synthesize ammonia from its elements using high temperature, high pressure, and osmium and uranium catalysts. Carl Bosch of BASF directed the conversion of Haber's bench reaction to a commercial process. The first ammonia factory was completed at Oppau near Ludwigshafen am Rhein in 1913. Instead of producing fertilizer it supplied the wartime Germany with nitric acid for explosives. The worldwide economic crisis of the 1930s and World War II postponed the take-off of large-scale fertilization until the early 1950s. Between 1950 and 1995 global synthesis of fertilizer ammonia rose from less than five million to about eighty million tonnes, and the application shares have shifted from more than nine-tenths in rich nations to about three-fifths in industrializing countries.

Fundamentals of the Haber-Bosch process have not changed, but its efficiency has increased substantially. The

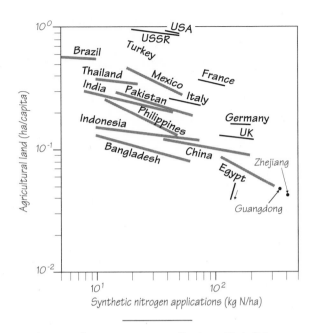

Increase of average nitrogen applications with declining availability of arable land between 1970 and 1990.

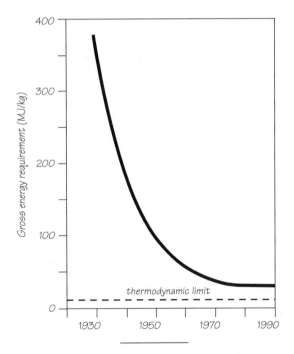

Declining gross energy costs of ammonia synthesis; thermodynamic limit of the reaction is 17.5 MJ/kg.

most important advances came during the 1960s when the replacement of reciprocating compressors by centrifugal machines cut the use of *electricity* by up to 95 percent, and led to larger and more economical plants. The almost universal shift from *coal* to *natural gas* as both the fuel and the feedstock (source of hydrogen) further lowered the energy cost of synthesis, while better catalysts increased the reaction yield. Total energy use in the best ammonia plants of the 1990s is below 35 MJ/kg N, less than one-fifth of the *coal* and *electricity* used by the first commercial syntheses of the early 1920s. The M. W. Kellogg Company of Houston is the leading provider of this efficient process: More than 150 of its large-scale plants were on stream or on order in 1995.

But ammonia is a gas under normal pressure, and hence it requires special transportation and application arrangements. That is why urea, a solid compound with the highest share of *nitrogen* (47 percent), has become the world's leading commercial fertilizer. Urea granules are easy to apply, no matter if the spreading is done by hand broadcasting onto small fields or by computer-controlled variable-rate spreaders in advanced precision farming. Additional syntheses needed to produce urea and other complex *nitrogen* fertilizers carry a major energy cost. Depending on the process and the age of the plant, production of urea needs at least 70 and up to 110 MJ/kg N, that of NH_4NO_3 between 60–100 MJ/kg N. Even so, synthetic nitrogenous fertilizers are great energy bargains. Assuming that they require on the average about 70 MJ/kg N, less than 2 percent of the world's primary energy consumption is used in their production.

This is a very small burden considering the nutrient's enormous existential importance. Careful assessments of *nitrogen* inputs during the mid-1990s add up to the annual flow of around 175 (extremes between 140 and 210) million tonnes of *nitrogen* into the world's croplands. Synthetic compounds supply about two-fifths of that total, and crop harvests remove about half of all inputs. With efficiency of fertilizer uptake ranging mostly between 45 and 55 percent this means that synthetic ammonia now provides nearly half of all *nitrogen* taken up by the world's crops. Because crops supply about three-fourths of all *nitrogen* in metabolized proteins (the rest comes from *oceans* and from grazing), at least every third person worldwide, and perhaps already two out of every five people, gets the protein in his or her diet from synthetic nitrogenous fertilizers.

Nuclear Weapons

Soon after he published his famous relativity paper, Albert Einstein began entertaining an "amusing and attractive thought." By 1907 the results of his cogitation were published. His conclusion was that "a physical system's inertial mass and energy content appear to be the same thing. An

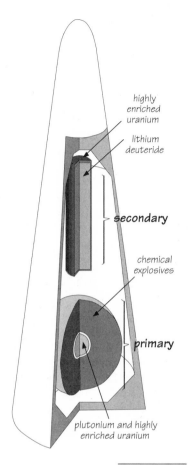

Typical thermonuclear warhead of a primary fission and a secondary fusion-fission component.

inertial mass is equivalent with an energy content mc²." Although Einstein could not demonstrate the law with chemical reactions—combustion of one kilogram of *fossil fuel* in oxygen reduces the mass of reactants by no more than 10^{-10}, too small an amount for him to measure—he knew that "for radioactive decay the quantity of free energy becomes enormous."

A generation later the discovery of the neutron (by Chadwick in 1932) was followed soon by the demonstration of nuclear *fission* (by Hahn, Strassman, and Meitner in 1938) and by the realization that a stunning amount of energy could be liberated in a chain reaction. Fissioning of one kilogram of ^{235}U diminishes its mass by one gram, releasing about 8.2 TJ of energy. The uncharted theoretical and engineering road to this release was traversed in an incredibly short time — and with unprecedented expenditures of energy, mostly for separating the fissile isotope of uranium.

The first nuclear test was at Alamogordo in New Mexico on July 11, 1945. Hiroshima was destroyed on August 6. The bomb explosion created temperature of several million degrees centigrade (conventional explosives reach only 5000°C) and the highest blast velocity of 440 meters per second. Its effect was equivalent to detonating 12.5 kilotonnes of TNT, with about half of its energy released in blast and a third as thermal radiation. These two effects caused tens of thousands of instant deaths, while the ionizing radiation caused both instant and delayed casualties. The Nagasaki bomb, exploded two days later, was nearly twice as powerful.

The postwar arms race between the two superpowers led to the design and testing of more powerful *fission* and

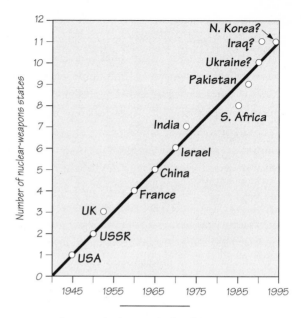

A new nation has acquired nuclear weapons roughly every five years since 1945.

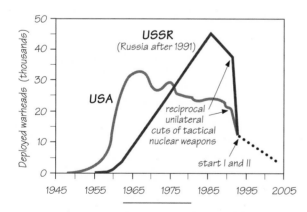

Rise and fall of superpower nuclear warhead arsenals. According to START treaties, the United States and Russia should each retain about five thousand strategic and tactical warheads.

fusion bombs (first tested in 1952 and 1953) as well as to the deployment of smaller battlefield *nuclear weapons*. The most powerful thermonuclear bomb tested by the USSR over Novaya Zemlya on October 30, 1961, was equivalent to 58 megatonnes of TNT — and less than fifteen months later Nikita Khrushchev revealed that the country had a 100-megatonne bomb.

When the two superpower arsenals reached their peaks during the mid-1980s, the two countries stockpiled nearly seventy thousand warheads. Just over a third of this total (some twenty-five thousand warheads) were strategic weapons (mostly between one hundred and one thousand kilotonnes per warhead) targeted at cities and military installations from land-based intercontinental or from submarine-launched ballistic missiles. *Fission reactors* powering the first U.S. submarines of the mid-1950s became the prototypes of commercial reactors widely installed in the country's new power plants after the mid-1960s, as well as those refined and serially produced for the aggressive French nuclear generation program.

The weapons did not remain confined to the two superpowers. Since 1945 a new nation has acquired nuclear capability roughly every five years: The United Kingdom during the 1950s, France and China in the 1960s, Israel and India in the 1970s, South Africa and Pakistan in the 1980s. Iraq and North Korea have come very close in the 1990s. Only three nations — South Africa and two successor states of the USSR (Belarus and Kazakhstan) — renounced their nuclear status, but at least three times as many either have the capacity to produce *nuclear weapons* and have so far abstained or are trying to acquire foreign help in launching domestic programs (Iran is the most prominent entry in the latter category).

During the decades of the Cold War many people were thinking about the unthinkable, first writing scenarios and then running computer models simulating the effects of either limited or full-scale thermonuclear exchange between the United States and the USSR. Even a limited Soviet strike, aimed just at the U.S. ICBM bases and with the combined power equivalent to a few thousand megatonnes, would have killed many millions of people. Launching of all strategic weapons by both sides during the mid-1980s would have unleashed explosive power equivalent to more than ten gigatonnes.

Of course, it makes sense to argue that it was precisely the incredibly destructive effect of *nuclear weapons,* one that would leave no winners, which has repeatedly pulled the superpowers from the edge of dangerous confrontations. *Nuclear weapons* may have been abominable, but effective, peacekeepers, but the price for this frightening benefit has been high.

Fixing the environmental damage done by decades of *nuclear weapons* production will cost hundreds of billions, and both the United States and Russia had weakened their economic well-being by investing trillions in an arms race producing ever more expensive but still only ephemeral deterrents. We will never know the exact figure, but the development and deployment of *nuclear weapons* and associated delivery systems has consumed at least one-tenth of all commercial energy used worldwide since 1945.

6

TRANSPORTATION AND INFORMATION

Growing shares of energy consumption in *fossil-fueled civilization* have been channeled into moving people and goods. The unprecedented degree of travel and trade are, together with the prodigious generation and exchange of *information,* the two key external attributes of high-energy societies. *Trains* were the first innovation making land *transportation* affordable on a mass scale, and *electricity* eventually pushed the speed of trains over two hundred and then three hundred kilometers per hour. *Steam engines* displaced wooden *sailships* before 1890, and *internal combustion engines* soon made an even greater difference.

Otto cycle engines powering passenger *cars* eventually created the largest manufacturing industry of the twentieth century and allowed a previously unimaginable degree of personal mobility. They also powered the first generations of commercial and military *airplanes.* Diesel engines proved invaluable on railroads, in shipping, and in heavy vehicles. The invention of *gas turbines* in the 1930s led to a rapid adoption of jet *airplanes* and the advent of mass international flying. These successive waves

of *transportation* revolutions had to be supported by mass deliveries of *crude oils* to refineries: expanding the capacities of *tankers* and *pipelines* met that challenge.

Telephones offered the first opportunity for instant remote communication, followed two generations later by a rapid diffusion of *radio and television. Computers* remained institutionalized underperformers for twenty years: only the invention of *microchips* turned them into personalized and powerful tools, and the still rising performance of chips is rapidly conflating all means of electronic *information* devices into new hybrid species.

Trains

Prime movers adapt readily to rails. *Horses* pulled wagons on first wooden railways; *steam engines* created the first fast, reliable, comfortable, and inexpensive land links; high-compression *internal combustion engines* (diesels) generating *electricity* for traction motors offered higher power at low cost; *electric motors* energized from overhead *transmission* lines revolutionized railways with unprecedented speeds. And the progression continues, as

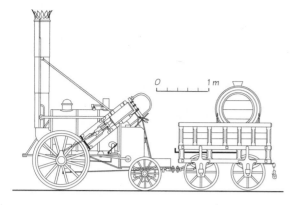

George Stephenson's 1829 *Rocket* embodied basic
ingredients of subsequent locomotive design.

Central Pacific's *229* models, an American classic first built
in Sacramento in 1882, served until the 1930s.

magnetic levitation now elevates trains onto a new plane
of performance.

The adoption of ***steam engines*** for locomotion was
delayed by the extension of Watt's patent, but even after
the patent's expiry the progress was slow. The first public
train service, connecting Liverpool and Manchester, began only in 1830, twenty-six years after Richard Trevi-
thick's experiments with a simple locomotive. The first
train on the pioneering railway was pulled by George Stephenson's locomotive *Rocket,* which won that honor in a
competition where it averaged just over twenty kilometers

per hour when pulling a load equal to three times its own
weight. This was much faster than the fastest carriage
pulled by a team of powerful ***horses***—but after just another fifteen years of locomotive innovation it seemed to
be a sluggish pace.

A maximum speed of a mile a minute (ninety-six kilometers per hour) was recorded on a scheduled English
railway in 1847. Coincidentally, that was also the year of
the most intensive railway-building activity in the kingdom. Shortly afterward the total length of American railways surpassed the British total, and the disparity grew
rapidly with the laying of transcontinental lines. The first
of these links was completed in 1869, and four were in
service by 1900.

By 1900 the Russians had been laying track for nearly
a decade across southern Siberia, but that transcontinental
line was not completed until 1917. All Western, Central,

With maxima around 220 km/h, Japan's *shinkansen,*
running on the Tokaido line since 1964, was the
first modern high-speed electric train.

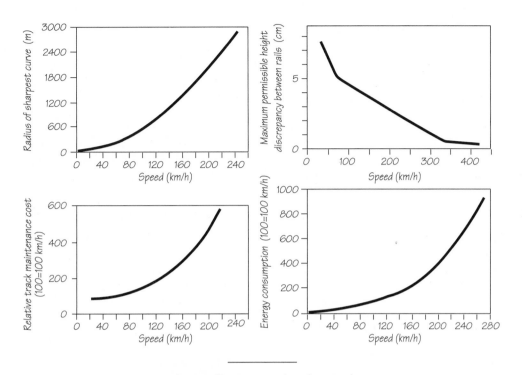

Higher speeds impose a number of engineering
constraints and require higher energy inputs.

and Northern European countries acquired a dense railway network by the beginning of the twentieth century, but the only countries with relatively extensive networks in Asia were India and Japan, and in Latin America only Argentina.

This rapid railway expansion engendered a great variety of technical innovations and enormous advances in reliability and comfort. By the middle of the nineteenth century passengers had a choice ranging from hard seats in unpartitioned and unadorned interiors to extravagantly appointed private compartments. Sleeping compartments were introduced on long-distance intercity lines (the first Pullmans were put into operation in 1864), while commuter *trains,* some of them descending underground (first in London in 1853) made the first wave of subur-

banization possible. Specialized freight cars were soon designed to carry bulk dry and liquid commodities and also refrigerated meat.

Technical improvements in *steam engines* continued during the first decades of the twentieth century, but the absolute gains in typical efficiencies and average speeds remained limited. Although the operating pressures of the best engines rose about fivefold in a hundred years, typical locomotive *steam engine* efficiencies still did not surpass 5 percent, and average speeds, even at main lines, remained below one hundred kilometers per hour. But on good tracks the fastest steam locomotives with streamlined cowlings reached speeds of two hundred kilometers per hour during the 1930s. After World War II railways faced increasing competition from trucking and also from *pipelines.* These two forms of long-distance *transportation* reduced the importance of freight *trains,* and the massive ownership of private *cars* and more affordable flying by jet *airplanes* weakened, and in North America nearly destroyed, both the commuter and long-distance railways.

But Japan and France have proved that in regions of high population density *trains* have an assured future. Japanese *shinkansen* begun operating in 1964, and its fastest *trains* (*hikari* and *nozomi*) reach maxima of, respectively, 220 and 270 kilometers per hour. French *trains à grand vitesse* (the first between Paris and Lyon in 1983) can reach almost 300 kilometers per hour, and in 1990 a modified unit of the train's latest version set the world record for a machine on rails at 515 kilometers per hour, the speed of a slow-flying jet. All modern fast *trains* are driven by *electric motors.* The first designs had DC supply at 1.5 kV, while now rectifiers (and regenerative brakes) are used with AC at 15–25 kV.

Freight *trains* regained competitiveness by moving much larger loads. Average capacity of freight cars grew from traditional ten-tonne boxcars to grain cars with capacities of forty to sixty tonnes, and to twice as large lightweight, permanently coupled hopper cars moving coal from mines to large power plants. These unit *trains,* made up of up to a hundred cars, pulled by groups of powerful locomotives and automatically unloaded in slow motion, have maximum capacities just over ten thousand tonnes.

Ships

The first *steam engines,* the inefficient Newcomen machines of the first half of the eighteenth century, posed no threat to *sailships.* But Watt's improvements made it possible to begin trials with steam-powered shipping: the massiveness of the early *steam engines* was much less of a problem on water than on land. The first commercial breakthroughs came with Patrick Miller's *Charlotte Dundas*

Two early paddlewheel steamers, Fulton's *Clermont* (1807) serving on the Hudson River, and the still fully rigged *Atlantic* of the Collins Line (1849).

in England in 1802 and Robert Fulton's *Clermont* in the United States in 1807, both small river-going boats propelled by paddle wheels. Many larger boats, both for inland shipping and for coastal runs, were built during the next decades, but the first crossing of the Atlantic came only in 1833, when *Royal William* sailed from Quebec to London. If need be it could really sail, inasmuch as it was still a fully rigged *sailship* with a *steam engine* aboard.

This crossing dispelled the long-held doubts that a *steamship* would not be able to carry enough fuel for such a long journey (the same concerns greeted early *airplanes* seven decades later). The first westward crossing of the Atlantic in 1838 was a race between *Sirius* and *Great Western* in 1838, but the era of these pioneering paddlewheels was shortlived as the first screw-propelled *ships* were launched during the 1840s. With better engines and better propellers speeds rose rapidly. During the late 1830s steamships could not beat *sailships* in a brisk wind. A decade later best steamships cut the transatlantic crossing to less than ten days, and before the end of the century that journey was normally scheduled to take less than six days. Their sustained speeds, averaging over ten meters per second, could not be matched by even the fastest clippers.

While wood limited the length of hulls to about a hundred meters, *steel* made it possible to build much larger *ships* and to design a variety of specialized vessels. In naval design this change produced the heavily armored, yet fast, dreadnoughts (named after British ship *Dreadnought,* launched in 1906) that were prominent in the Atlantic battles of World War I. They were also among the first *ships* to be powered by *steam turbines,* which were

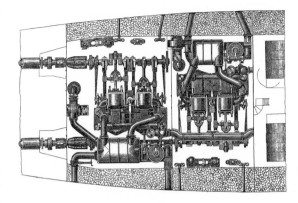

Engine room with two large horizontal steam engines on HMS *Iris* (1878).

much more efficient and much more powerful prime movers than *steam engines.* But their primacy lasted only two or three decades: by the 1930s diesel engines became the leading marine prime movers for both naval and commercial *ships.*

Larger passenger vessels made it much easier — although hardly very pleasant for steerage passengers — to transport tens of millions of European migrants to the New World and to Australia. The era of increasingly larger and faster *ships* came to an abrupt end, however, when *gas turbines* began to power the first commercial jet *airplanes.* Just five years after the launching of the fastest passenger vessel (the *United States* in 1952), transatlantic airliners carried more people than did *ships.*

At the same time, cargo (merchant) *ships* were experiencing rapid increases in both numbers and capacities. They became specialized along three basic lines. Heavily traded raw materials and basic commodities are now moved by bulk carriers; *tankers* are a special category carrying *crude oils,* liquefied *natural gases,* and liquid industrial chemicals; dry loads include mostly mineral ores, *coal,* and *grains.* Automated load handling is essential: the largest dry bulk carriers now surpass 100,000 dwt.

Majestic of the White Star Line (formerly the German *Bismarck*), one of a few giant liners of the early twentieth century.

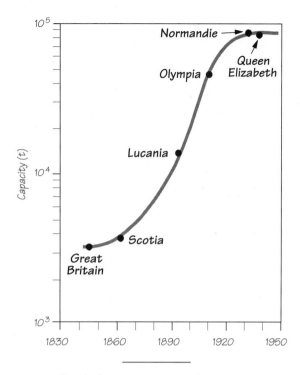

Growth of maximum passenger ship tonnages.

Container *ships* efficiently move a growing assortment of manufactured goods to major ports where trucks take over the final distribution. Finally, there is a great variety of specialized merchant vessels carrying vehicles and *trains* (conveniently as roll-on/roll-off ferries), and products that could be neither loaded in bulk nor containerized (such as tens of millions of *cars* from Japan and Europe and millions of live sheep from Australia to the Middle East).

The two most important naval innovations during the twentieth century have been long-range submarines and aircraft carriers. Diesel-powered submarines were first used with devastating effects by Germany during World War I, and again during World War II. *Fission reactors,* developed rapidly after World War II (the U.S.S. Nautilus was launched in April 1954), made submarines equipped with *nuclear weapons* a critical component of superpower

strategic balance. The first aircraft carriers were launched in the early 1920s, and their successors were engaged in critical Pacific battles during World War II, accommodated jet *airplanes* after 1950, and a number of them also became powered by *fission reactors.* In spite of some enthusiastic early expectations, nuclear propulsion has not made any inroads in commercial shipping.

At the close of the twentieth century great naval fleets are shrinking once again, reflecting the easing of the long superpower contest. Fishing vessels are increasingly idle because of the overexploitation of some of the world's richest fishing grounds. Merchant shipping grows in order to accommodate the continuing growth of global trade — but, ferries aside, passenger shipping is now overwhelmingly limited to recreational cruising. Several new cruise *ships* are now as large as the largest transatlantic liners operating between 1920 and 1960.

Bicycles

The late appearance of the most efficient mode of human-powered locomotion and its subsequent slow and retrogressive development are puzzling. Modern, comfortable, and relatively safe machines were designed only during the 1880s, the decade when *cars* began their still unfinished

Starley's 1886 safety bicycle, the precursor of modern machines.

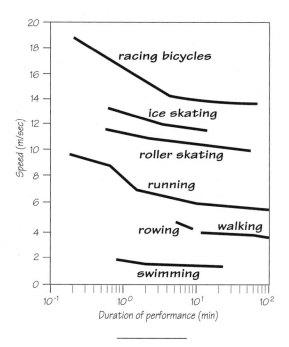

Cycling is the fastest, as well as the least energy-demanding, form of human-powered locomotion.

ascent. Between the design of essentially modern *bicycles* and the demonstration of their first steerable precursor lie about three generations of only partially explicable false starts and tentative advances.

In 1817 Karl von Drais convincingly demonstrated that with his machine—propelled much like contemporary hobbyhorses by kicking but, unlike them, having the advantage of a steerable front wheel that made balancing possible—he could outrace not only runners but even a *horse*-drawn carriage. Drais had many imitators, but it was only in 1839 that a Scottish blacksmith named Kirkpatrick Macmillan built the first true steerable *bicycle* by connecting rear-wheel cranks with rods to swinging pedals. His innovation led nowhere, perhaps because of the contemporary preoccupation with rapidly advancing travel by *trains.*

A revival came in 1861 with Pierre Michaux's *vélocipède,* a true *bicycle* with pedals attached directly to the front wheel. Adherence to this design led almost inevitably to machines with large front wheels: their enlargement to diameters of nearly two meters was the simplest way to achieve higher speeds by getting a longer distance for every pedal revolution. Of course, it was also the way to more frequent accidents—and it excluded not only any cautious would-be adult rider, but also all children, short men, and, given that era's dress requirements, most women.

A great deal of tinkering was done during the 1870s with asymmetrical *bicycles,* but their demise was rapid once the real thing arrived in the mid-1880s, when John Kemp Starley and William Sutton introduced designs that, in their final form, incorporated the essential features of modern *bicycles:* equal wheels, chain-and-sprocket drive to the rear wheel, a diamond frame of tubular steel, pneumatic tires (after John Dunlop's 1888 patent), and a back-pedal brake (after 1889). The only fundamental improvement of the 1890s was the addition of multiple gears changed by derailleurs.

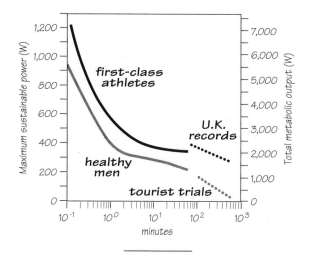

Maximum sustainable power and total metabolic output in cycling.

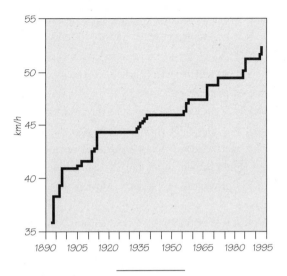

The longest distance covered by cycling in one hour rose from just over 35 km in the early 1890s to over 50 km a century later.

Bicycles — soon also equipped with lights, load carriers, and tandem seats — finally became the most widely accessible means of personal locomotion for commuting, shopping, recreation, or competitive racing. But by the 1920s in North America and after 1950 in Europe they had to face a new form of enticing competition from increasingly more affordable *cars*. Consequently, they had their greatest post–World War II success in Asia, particularly in China, which remains the world's largest producer of *bicycles* (over thirty million annually since 1985). The conjunction of environmental concerns, a fitness vogue, and aggressive marketing of new machines (mountain bikes) brought about the resurgence of *bicycle* sales in rich countries.

Regardless of the length of a race, cycling is the fastest mode of human-powered locomotion. Deployment of the largest leg muscles maximizes conversion efficiency of metabolic energy into rotary motion and allows the cyclists to achieve the highest possible human power outputs. Naturally, the duration of peak power outputs declines exponentially with rising performance. Bursts in excess of 1.1 kilowatts are sustainable for just a few seconds, professional racers can sustain rates above four hundred watts for an hour, while exertions lasting more than one hour are limited to less than two hundred watts in untrained men. The best sustained performances are achieved with forty to fifty rotations per minute, with saddles raised four to five centimeters above the normal height, and with longer-than-normal pedal cranks coupled with higher-ratio gearing.

Anaerobic conversions powering short bursts are only about 10–13 percent efficient, and so the highest efficiencies, just over 20 percent, will be reached by elite aerobic performers working at peak kinetic power outputs. Improved *bicycles* (most notably lighter frames, made of **aluminum** or titanium alloys and composite materials), better tires, and also better riders' equipment (sleek suits, streamlined helmets) and new riding positions (notably the "egg," with the shoulders resting on the back of the hands, which grip narrow straight handlebars) help to minimize friction and to set new speed records. The world's best riders against the clock can now cover more than fifty kilometers in one hour.

Cycling is not only the fastest means of human-powered locomotion on land but also on water, and it is the only energy conversion allowing humans to fly. *Flying Fish II,* a watercraft pedaled like a bicycle, reached about 6.5 meters per second, faster than a single rower. *Gossamer Condor,* the first human-powered aircraft to fly around a one-mile, figure-eight course in 1977, averaged 4 meters per second; in 1984 the cyclist inside MIT's *Monarch B* completed a triangular 1,500-meter course at nearly ten meters per second.

Cars

The historical verdict on passenger *car,* as it moves into the second century of its existence, remains uncertain. The

The Ford Model T (1908), the first mass-produced car.

impact of this machine — which was invented in Europe, elevated to a way of life in North America, and transformed into an essential ingredient of global *embourgeoisement* — has ranged from liberating to confining, from indisputably beneficial to obviously harmful. Freedom of personal movement gained with private *cars* has been so addictive because, in Kenneth Boulding's felicitous analogy, the ownership of this mechanical steed turned even the humblest driver into a knight with aristocrat's mobility looking down at pedestrian peasants.

The personal and professional mobility conferred by *cars* has been among the most powerful social forces of the twentieth-century Western world. The freedom of individual movement conferred by *cars* is extremely appealing — but its cost is rather steep. Although most North American adults spend at least 250 hours a year driving a *car*, they have to spend 350 hours, or nearly 20 percent of their working time, to earn money for the vehicle's purchase and for its insurance, fuel, and repairs.

Economists would point out the enormous contribution of *cars* to the higher standard of living. Carmaking became America's leading industry in terms of total product value during the mid-1920, a primacy replicated in every major Western economy after World War II. Huge segments of other leading industries, ranging from **steel** and rubber to aluminum and plastics, depend on carmaking, and substantial shares of other industrial and service activities, ranging from catalytic oil refining to skiing resorts, would not exist without *cars*. Nor would, obviously, suburbs (now extending into exurbs) and multilane highways: Their construction has absorbed huge amounts of capital in every rich country, and even larger sums will be required for their upkeep.

Environmentalists would focus on the car's enormous air pollution impacts, and on its destructive reordering of urban space. Large concentrations of *cars* emitting nitrogen oxides, carbon monoxide, and hydrocarbons have brought often acutely damaging concentrations of photochemical smog to most large metropolitan areas. Overall long-term costs (mainly health and crop damage) of this pollution may be as large as the immediate toll of accidental injuries and deaths. Highly traveled roads and freeways have destroyed or downgraded neighborhoods and brought in incessant noise and pollution, all too

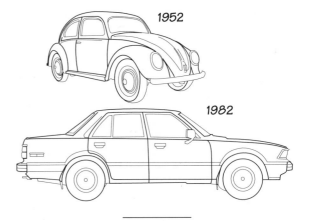

Two exports that changed the global car market: Germany's Volkswagen, the world's most successful small car (a 1952 model shown here), and Japan's Honda Accord (1982 model), an epitome of high-quality Japanese carmaking.

frequently without increasing typical commuting speeds. And new rapid *trains* are clearly superior for comfortable intercity travel.

Cars pollute so much because so little energy converted by *internal combustion engines* ends up doing the useful work of locomotion. A well-tuned engine converts just over 30 percent of gasoline into rotary motion, but the frictional losses during driving lower this by about 20 percent. Partial load factors, inevitable during the urban driving, reduce this by another 20 percent. Accessory loss and automatic transmission may nearly halve the remainder so that the actual efficiency is no more than 10–12, and often as low as 7–8 percent.

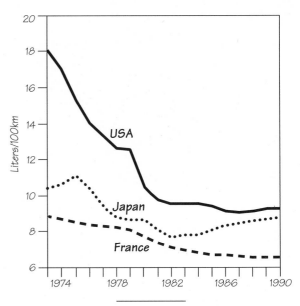

Average fuel consumption of new cars, 1973–1990.

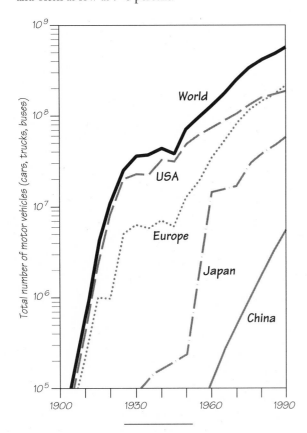

Global rise of vehicle registrations.

Low conversion efficiency combined with still rapidly rising registrations means that *cars* have been claiming ever higher shares of global primary energy use. When the assembly of the first truly mass-produced car, Ford's famous Model T, was discontinued in 1926, there were fewer than 25 million privately owned vehicles worldwide (nearly three-quarters of them in North America), consuming less than one-tenth of the world's output of *crude oils.* In 1995 worldwide passenger-*car* registration surpassed the half-billion mark, and the global rate approached eleven people per car. The rate fell below 1.8 in the United States, and it was not far behind in Germany (1.9) and Japan (2.1). Gasolines burned in North American vehicles amounted to two-fifths of all liquid fuels consumed in the continent, and the share for all rich nations was almost exactly one-third of all *crude oils.*

Simple calculations show that extending the rich world's rate of car ownership to the world's most populous nations could not be done with existing vehicles:

China alone (with 1.2 billion people) would have more passenger *cars* than the whole world had in 1995, and global crude oil production would have to more than triple to fuel some three billion *cars*. But we cannot be surprised that modernizing countries succumb to the same addiction we have been nurturing for a century.

The only rational way ahead, for rich and poor countries alike, is to drive radically new *cars*. Possible improvements are impressive. In the early 1970s the efficiency of new U.S.-made *cars* was below fifteen miles per gallon; a generation later it rose to nearly thirty, and the best *cars* on the road reach double that rate. The best available designs go close to one hundred, and two hundred miles per gallon should be possible in a decade or two. These lightweight, streamlined vehicles, built more like *airplanes,* most likely would be hybrid *cars,* where various prime movers would power *electric motors* directly driving the wheels.

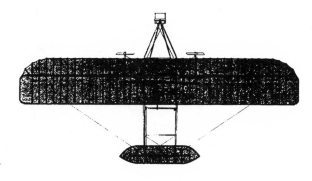

Plan and front views of Wright brothers' 1903 plane; its wing span was 12.29 meters.

Airplanes

The first successful flight was an extraordinary achievement of just two enterprising individuals: the Wright brothers had almost no precepts to follow. First they tackled the fundamental challenges of aerodynamic design and its flight testing. They solved the two crucial problems of balance and control and proper wing design with their 1902 glider. Then they built two wooden propellers driven in opposite direction by *bicycle*-chain geared drivers.

They cast a light four-cylinder engine with an *aluminum* body and a *steel* crankshaft. This machine was rated at eight horsepower, but in tests and flights it actually delivered about 50 percent more power. They accomplished the first successful flights of a heavier-than-air *airplane* on December 17, 1903, over the seashore sands of Kill Devil Hills, North Carolina.

The subsequent development of *airplanes* was remarkably rapid. *Airplanes,* both fighters and bombers,

became increasingly common over battlefields and cities during World War I. The first scheduled international link, the daily London–Paris flight, came just after the war in 1919. By the late 1920s lightweight *aluminum* bodies started to displace wood and canvas, and ten years later large hydroplanes made regular transoceanic flights possible.

Internal combustion engine used in *airplanes* improved very rapidly, supporting both faster speeds and longer ranges. Those powering Boeing's 1936 Clipper were about 130 times more powerful than Wright's 1903 machine whose weight/power ratio was more than ten times as high. Further substantial improvements came with the demands of World War II.

The aircraft industry was revolutionized by introduction of mass-production techniques, by the design of highly maneuverable fighters and heavy long-distance bombers, and by the introduction of *gas turbines* for

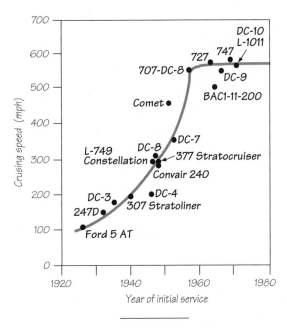

Leaving the limited service by the supersonic
Concorde aside, exponential increase of typical cruising
speeds of commercial planes reached a plateau with
the introduction of the Boeing 707 in 1957.

fighter **airplanes.** Flight control was radically improved
by universal adoption of radar.

The speed of sound was surpassed on October 14,
1947, with the Bell X-1 plane, and the rapid development
of military jets was soon followed by commercial designs.
The British Comet became the first passenger jet in 1952,
but it had to be withdrawn after a series of crashes caused
by structural defects in its fuselage. The Boeing 707,
introduced in 1958, thus became the first intercontinental jet.

Other Boeing designs became even more important.
The short- to medium-haul 737, introduced in 1967, has
become the best-selling jet of all time with order surpassing 2,500 **airplanes.** And the 747, the first wide-
bodied long-range jet, introduced in 1969, started the age

of mass intercontinental travel: the thousandth 747 was
delivered on October 12, 1993, and by that time the aircraft had flown 1.5 billion passengers.

All large passenger jets are powered by huge turbofan
gas turbines with more than 200 kN of thrust developing
up to 60 MW of power. The most powerful jet engine ever
built, the GE4/JSP, was to propel the supersonic Boeing
B2707–300SST, and at Mach 2.7 it would have delivered
248.32 MW. The cancellation of the SST program by the
U.S. Congress ended the prototype's development. Concorde, the only supersonic plane in service, which has
never been profitable, is powered by four 169.3 kN Rolls
Royce/SNECMA turbojets. Its maximum speed of Mach

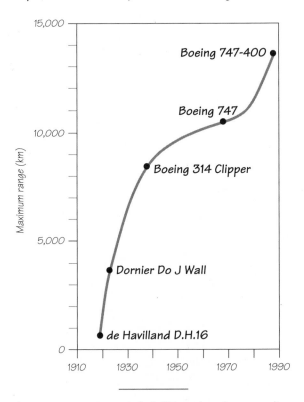

In contrast, improvements in fuel efficiency have been extending
the maximum flight range of commercial planes.

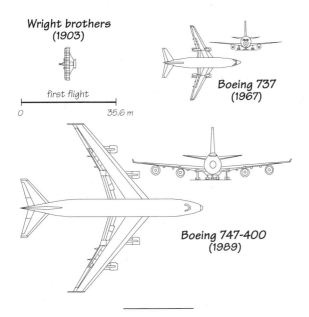

Wright brothers
(1903)

first flight

0   35.6 m

Boeing 737
(1967)

Boeing 747-400
(1989)

Plan and front views of the Boeing 737, the smallest and the most
successful of the company's family of commercial jets first flown
in 1967; and of Boeing 747–700, the latest upgrade of the
first jumbo jet launched in 1989. The Wright brothers' plane
and its first flight path are shown for comparison.

2 remains unrivaled in passenger *transportation,* but it is routinely surpassed by many combat aircraft.

The two aircraft producers dominating the global passenger jet market—the U.S. Boeing/McDonnell Douglas and European Airbus consortium—are building about thirty different types of *airplanes,* including various wide-body designs accommodating up to nearly five hundred people. Top cruising speeds of these *airplanes* now approach the speed of the sound, and the largest ones can fly well over ten thousand kilometers without refueling.

The combination of near-sonic speeds, growing ranges, and of the multitude of airlines and frequency of services has made it possible to reach any continent within a single day—and to do so with steadily declining real costs. Jet *airplanes* created the mass long-distance tourism to warm beaches and exotic destinations, opened up new trade opportunities (from cut flowers from Holland and Colombia to overnight mail deliveries), and contributed greatly to the rising integration of the global economy. The undesirable effects of this enormous mobility include international terrorism involving aircraft hijacking, widespread drug smuggling, and the rapid diffusion of infectious diseases.

In spite of enormous airline expansion, energy consumed by *airplanes* worldwide represents a fraction of fuel needed by *cars.* Jet fuel amounts to less than 6 percent of the world's refined products, and 11 percent of the U.S. refinery output. In contrast, gasolines accounted for 25 percent of the world's refined fuels and for about 45 percent of liquid-fuel consumption in the United States.

Rockets

The journey of Apollo 11 to the Moon began on July 16, 1969, with the two and a half minute burn of the liquid-fuel (kerosene and hydrogen) engines of the 109-meter tall Saturn C5 *rocket.* In order to impart the escape velocity of 11.1 kilometers per second to the mass of 43 tonnes this burn had to develop combined thrust of nearly 36 MN, an equivalent of about 2.6 GW. Among all other prime movers, only the largest *steam turbines* can deliver power of the same order of magnitude (about 1.5 GW, and can do so for thousands of hours), but no energy convertor can even remotely approach the low weight/power ratio of *rocket* engines designed to carry payloads, be they communication satellites or *nuclear weapons,* into space.

Even when all the fuel is included in the mass of large *rocket* boosters, their weight/power ratio would be just around 0.001 g/W, half of the rate for jet engines in the best fighter *airplanes,* and the thrust/weight ratio of

The ***ethanol***-powered V-2 missile had the maximum range of 340 kilometers and a destructive payload of one tonne. Its sea-level thrust of 249 kN and the maximum burn of less than seventy seconds gave it power equivalent to about 6.2 megawatts. Postwar U.S. research under the leadership of Wernher von Braun, and similar Russian efforts under Sergei Korolyov, began the intensive development of ***rocket*** engines marked by the launching of the first satellite in October 1957, first manned space flights

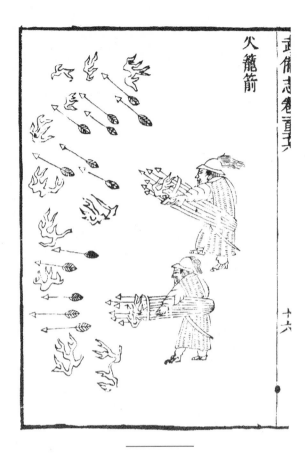

Launching of rocket-arrows by soldiers of the Ming dynasty.

Liftoff of Apollo 11.

individual ***rocket*** engines could be as high as 150, ten times better than in the topline jets.

What has been in retrospect a surprisingly short journey toward realizing such spectacular feats of propulsion begun in earnest only during the first decades of the twentieth century with design and testing of toylike experimental ***rockets*** by Robert Goddard and Hermann Oberth. Decisive advances came first with the wartime German development of ballistic missiles, which caused many British casualties and damage to housing but made no difference to the course of the war.

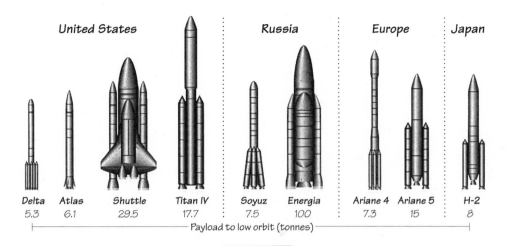

Launch rockets of the 1990s.

of the early 1960s, Moon landings between 1969 and 1972, and unmanned exploration of Mars and Venus and probes reaching the outer planets of the solar system.

The military, and hence also political and socioeconomic, impact of these novel prime movers was much more important. The superpower arms race produced a large variety of missiles ranging from those for short-range battlefield use to those capable to carry several nuclear warheads to separate targets overseas. These missiles, launched either from reinforced silos or from nuclear submarines and known in the acronymic argot as MIRVed ICBMs or SLBMs, eventually became the keystones of protracted and exhaustingly expensive strategic stalemate.

But it is also possible to argue that the availability of more powerful *rockets* able to carry heavier payloads has not had its greatest impact either in terms of space research or military balance, but because it literally launched a still far from finished revolution in global communication. In conjunction with growing powers of *computers* it made navigation, be it by *airplanes* or *ships,* much safer, it turned *telephones* reliable and inexpensive means of global communication, and it extended the reach of *radio and television* to entertain, to influence, and to create markets on previously unattainable scale.

Pipelines

Out of sight, out of mind. Few people outside the industry think daily about hydrocarbon *pipelines,* and the reason is not only that the lines are buried: their accidental breaks are so rare that they, unlike the much more accident-prone *tankers,* hardly ever make headlines. This makes them the most environment-friendly mode of continental *transportation,* an appeal further strengthened by the fact that they are also very compact—one m-diameter line can carry fifty million tonnes of oil a year—and very quiet. And because they are also by far the least expensive way of moving huge volumes of liquids and gases they are superior on all important counts to all other means of bulk *transportation* on land.

Three conditions had to be satisfied first to make them so: excellent steels, mechanized laying, and reliable pumps. Reliable large-diameter, high-pressure lines would

Mechanized laying of a large-diameter natural gas pipeline.

long-distance oil line was built in Pennsylvania (180 km between Bradford and Williamsport in 1879), the first ones crossing a country in 1931 from Texas to Detroit, soon followed by two longer lines (2200 and 2400 km) to the East Coast. Large post–World War II *natural gas* discoveries in Alberta and in European Russia led to the building of the world's longest pipeline, the 7200-kilometers long TransCanada, and to a rapid development of extensive Soviet network. Outside of Romania, post–World War II Europe had only some two hundred kilometers of *pipeline*—but by 1970 its soaring oil demand had brought more than ten thousand kilometers of lines with capacity of over 400 million tonnes.

The search for hydrocarbons in remote regions and in offshore basins, a trend accelerated after 1973 by OPEC's oil-price increases, resulted in new construction challenges as *pipelines* were laid in the notoriously stormy waters of the North Sea and in the high Arctic. Permafrost in Alaska and Siberia necessitates aboveground mounting on well-anchored supports, but these obstacles were successfully overcome with the construction of America's most expensive Transalaska oil *pipeline* from the North Slope to Valdez, and with the completion of the world's longest system of *natural gas* lines extending some 6500 kilometers between supergiant fields in Western Siberia and European Russia, the Central Europe, and Germany and Italy with pipes of 2.4 and 1.4 meters in diameter.

Because of the extensive West European gas network, which is also connected across the Sicilian Channel to North African deliveries, the continent is now spanned by gas lines from Sicily to Norway and from Spain to the Urals. Enormous gas fields in Iran and Saudi Arabia are closer to European markets than Western Siberia, but chronic political instability in the region makes it unlikely that they will become the starting points of large-diameter *pipelines* in the near future.

be impossible without seamless *steel* pipes (now available in length up to twenty-four meters) protected against corrosion and without *gas turbines* powering efficient pumps. And *pipeline* construction has evolved into a marvel of mechanization and automation, from trenching and delivery, cleaning and coating of pipes to their welding, testing, laying, and burial.

Predictably, record-breaking construction of *crude oil* and *natural gas pipelines* followed the discoveries and developments in major hydrocarbon basins, as well as the rising import needs of rich importing countries. The first

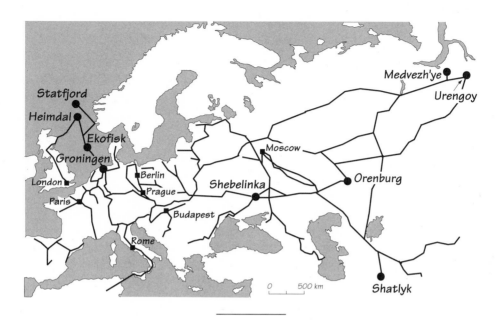

The European network of natural gas pipelines extends
from Western Siberia to the North Sea.

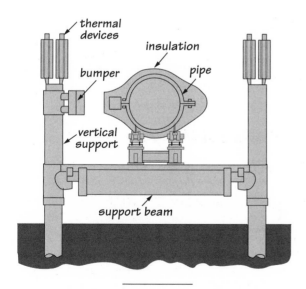

Elevated pipelines are common in permafrost
areas of Siberia and North America.

Tankers

Development of *tankers* is a nearly perfect case of effective technical response to economic pressures, to the necessity to transport increasing volumes of *crude oils.* And it is also a fine illustration of how nontechnical factors can suddenly limit an exponential expansion. The first *tanker, Gluckauf* launched in 1884, had a capacity of just over two thousand tonnes, and most of the pre-World War I oil was carried by vessels of just around ten thousand tonnes. A *tanker* surpassing twenty thousand tonnes was launched in 1921, but this was an isolated exception the early 1920s as oil trade stagnated during the next two decades. World War II, most notably the unprecedented needs to fuel the U.S. invasion of Europe and the naval reconquest of the Pacific, led to a mass construction of T-2 *tankers* carrying sixteen thousand tonnes.

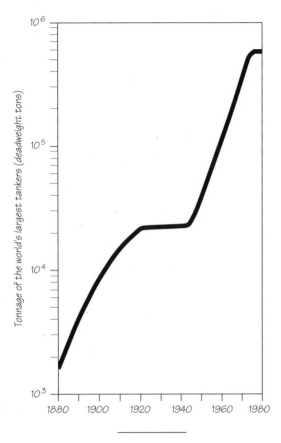

Growth of maximum tanker capacities, 1860–1980.

depth limited *tankers* to about fifty thousand tonnes) after its temporary closure in 1956.

Supertanker label begun to shift rapidly to a succession of larger *ships.* By the mid-1950s it referred to *tankers* carrying fifty thousand tonnes. A decade later, when it described vessels with more than a 100,000-tonne capacity, trade magazines begun referring to mammoth (over 200,000 tonne) and supermammoth (over 300,000 tonne) *ships* that were at that time either under construction or in the planning stage. Shipping *crude oils* on these huge vessels virtually eliminated distance as a variable in purchasing decisions. With cheap diesel fuel, small crews (fewer than twenty people), and rapid turn-arounds (seventeen days from the Persian Gulf to Europe) supertankers made oil a perfectly global commodity, deliverable from anywhere.

Trend-lines were pointing inexorably to a million-tonne ship — but the trend stopped rather suddenly at just over 500,000 tonnes. A slight decline in global *crude oil* trade was only a partial reason. There was no insuperable technical problem to build a *tanker* with twice the capacity, but even the largest existing *ships* began to run in into a combination of disadvantages undercutting their in any

These decommissioned *ships* became the first postwar commercial supertankers, but in less than a decade their capacities seemed puny. Five factors were behind the rapid post-1950 growth of *tanker* capacities. The two general reasons were large increases in global *crude oil* consumption and the quest to reduce *transportation* costs; they alone would have favored considerable economies of scale offered by larger *ships.* Three time-specific reasons were shifting pattern of world oil trade, above all the emergence of Japan as a large importer; the replacement of the Caribbean by the Middle East as the major source of European imports; and a tendency to avoid the Suez Canal (whose

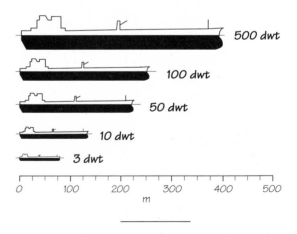

Dimensions of some typical small, medium, and large tankers.

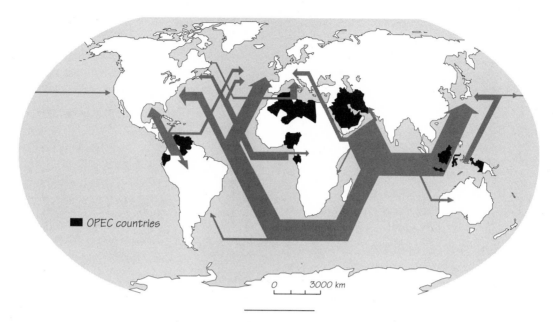

Relative importance of major crude oil export streams
carried by tankers from OPEC countries.

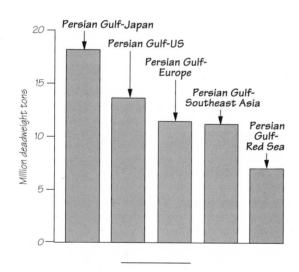

Supertankers on main shipping routes.

case diminishing economies of scale. They can call on a very limited number of ports, they cannot navigate in shallower waters or in constrained channels, they require long distances to stop, and their huge loads pose risks of enormous accidental oil spills and consequent high costs of insurance and eventual cleanup and litigation.

Shipments of liquefied *natural gases* (at −162°C) in *tankers* are inevitably much more expensive. They began in the early 1960s, and now include deliveries to Western Europe, the United States, and Japan. Fortunately, there has never been an LNG *tanker* accident. The explosion of a fully loaded ship would release energy comparable to detonation of a *nuclear weapon* equivalent to ten Hiroshima bombs.

Telephones

Well into its second century of existence, the *telephone*'s allure remains undiminished. More than that, the latest version of the second oldest mode of *electricity*-borne telecommunication—a progressively smaller cellular phone—is reconquering countries that have been saturated by the traditional device for many years. This is a unique development among well-established electronic devices.

The telegraph, the first means of wire-borne transmission, seemed for long so indispensable, its extension by undersea cables so remarkable. For a hundred years after the late 1830s, when William Cooke and Charles Wheatstone patented its first reliable version and when Samuel Morse designed his famous codes, it dominated the intercontinental flow of *information.* Although the first American transcontinental *telephone* link came in 1915, the first transatlantic *telephone* cable was laid only in 1956 (the radio-phone was in use since the late 1920s, but it was neither reliable nor cheap). Now for most people alive the telegraph is only a historic artifact.

Even *radio and television* are encountering strong competition from a new species of personal *computers*—but *telephones,* with melodious chiming or gentle chirping

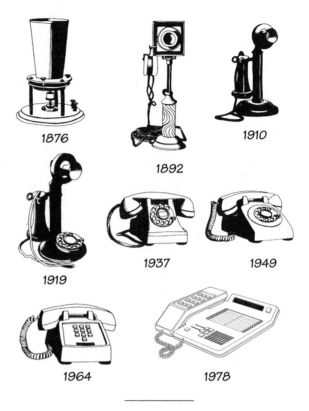

A century of telephones. Bell's original transmitter (1876), two early desk sets (1892 and 1910), the first dial telephone (1919), standard "300" (1937) and "500" (1949) desk sets, and one of the first electronic phones (1978).

replacing harsh ringing, grow still more ubiquitous. They owe their resurgence to a combination of *rockets* and *microchips.*

The classic *telephone*—patented in one of the strangest episodes in the history of invention by Alexander Graham Bell on February 14, 1876, just hours ahead of Elisha Grey's independent filing—was a rather simple electromechanical device based on principles predating its invention. Only slightly improved versions of Bell's model were put rapidly into local and regional service, and basic phone design remained surprisingly conservative for decades.

A. G. Bell's conceptual sketch of the first telephone: Its receiver was a tuned reed.

This inertia had nothing to do with possible higher energy needs of better systems. **Telephones** use very little **electricity,** which they receive from a central battery system backed up by emergency generators; this arrangement assures that phone lines remain open even during extended **electricity** outages. By far the most important reason for this slow progress was the large investment of monopolistic companies in the installed equipment, which they expected to last for decades.

Rotary dialing came only a quarter century after Bell's invention, and the first one-piece set incorporating a transmitter and receiver in the same unit was introduced only in the late 1920s. Then, with the exception of push-button dialing (introduced in 1963, and requiring rather complicated wiring), **telephone** design stagnated until the beginning of a swift conversion to electronic telephony begun in the late 1970s.

Electronic phones finally took the advantage of rapidly advancing **microchip** design to incorporate much simpler push-button dialing, to replace carbon microphone by the dynamic transducer, and to substitute tiny integrated-circuit components for the bulky bell ringer, transformer, and oscillator. Phones got smaller, lighter, and cheaper.

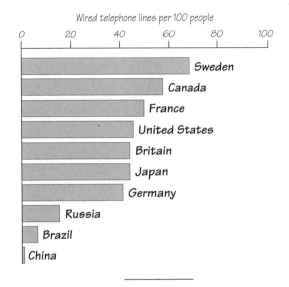

Wired telephone lines per hundred people in the early 1990s.

Much more affordable intercontinental service became possible with the launching of commercial telecommunication satellites beginning with *Early Bird* in 1965. Placed in geostationary orbits, these satellites can simultaneously carry tens of thousands of calls as well as **radio and television** broadcasts. The replacement of **copper** wires by optical fibers, carrying many more signals as pulses of light, was the third key component of this great innovation process.

And the transformation of traditional wired **telephones** was followed closely by the ascent of cellular phones. Residential cordless phones, first available during the mid-1970s, were the first tiny step in that direction. Between 1980 and the mid-1990s **microchip**-driven miniaturization shrank the internal volume of the cellular **telephone** by a factor of one thousand; in a matter of years those bulky boxes installed as prestige items in expensive **cars** became slim, foldable contraptions fitting into a shirt pocket — and with tumbling prices also filtering down into decidedly modest income categories.

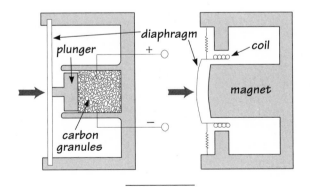

Carbon transducer, invented in 1876 and used until the 1970s, and the dynamic transducer used in electronic phones.

The worldwide growth of cellular *telephone* ownership has been much faster than originally forecast — and there is a huge potential both in terms of new customers and new services, including replacement of analog signals by superior digital telephony. In poor countries in general, and those with rugged terrain in particular, it is much cheaper to establish cellular networks than traditional wired service. And the coming global cellphone networks will do away with wires and fibres by transmitting signals via fleets of satellites launched into low, intermediate, or geostationary orbit, a change assuring the lasting indispensability of *telephones* in the twenty-first century.

Radio and Television

Radio, the first mode of wireless communication, is clearly past its prime. Although both AM and FM stations are crowding the available spectrum, often with nonstop programming, today's *radio* broadcasts provide most of the time just an aural background for other activities rather than a focus of undivided attention. This retreat began almost as soon as *radio* reached its peak influence: the birth, rise, and decline of the medium has been compressed into a single lifetime, and those who listened to first broadcasts during the 1920s can still readily recall their genuinely riveting effect of immediacy and bridged distance.

Radio's preliminary beginnings were in the 1890s with Marconi's increasingly impressive experiments with wireless telegraphy. The following decade brought inventions of two devices that made practical receivers possible — the diode by J. A. Fleming, and the triode by Lee de Forest. But mass production of *radios* and diffusion of their ownership began only after World War I, and it proceeded very rapidly: just two years after the beginning of first commercial broadcasts in Pittsburgh the United States had one million listeners.

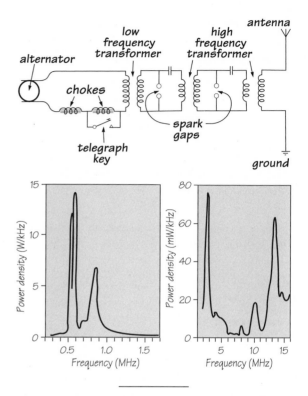

Marconi's Poldhu transmitter and spectra of its low and high frequencies. Total energy radiated was about 1.8 kW, and the average received at St. John's was 50 pW.

By the late 1930s *radio* was a commonplace family possession throughout the industrialized world, and the invention of the transistor (by William Shockley, John Bardeen, and Walter Brattain in 1947) led to its widespread adoption in Latin America, Asia, and Africa. After 1970, in what might be called the third diffusion wave, tiny portable *radios* equipped with earphones (and often coupled with cassette players) found many buyers among the people on the move (be they commuters or joggers) in the otherwise saturated market.

Widespread diffusion of *television* came after a much greater delay. The key ingredient — cathode ray oscilloscope—was invented by Karl Braun in 1897, but the first experimental images were transmitted only thirty years later, and the bulky black-and-white sets became common in the United States only during the 1950s and in Europe a decade later. Mass ownership was achieved much faster in many poor countries once relatively cheap color sets came on the market in the 1980s.

But broadcast *television* is in trouble. Its rapid diffusion could not eliminate movies, and now a highly diversified and seemingly unlimited flow of visual opiate is being administered by cable transmissions, by VCRs replaying movies and games, and by *computers* connected to CD ROMs or roaming the WorldWideWeb with browsers.

Plugged-in *radios,* now commonly bought as a part of compact audio systems, are relatively low consumers of *electricity,* needing less power (mostly just 10–30 W) than any other common household appliance except for the electric clock (typically just a few watts). At fifty to a hundred watts, color *televisions* now rate only half as

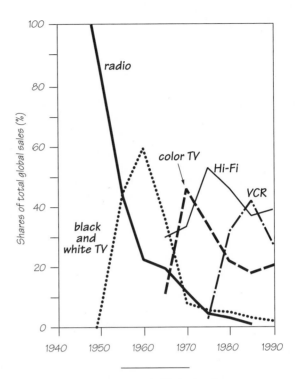

Successive waves of global sales of major consumer electronics products.

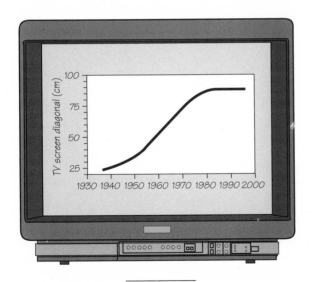

Growth curve of television screens (diagonal in cm).

much as a generation ago—but standby demand by switched-off, remote-ready sets (averaging 4 W), by cable TV boxes (about 10 W), and by VCRs (5–6 W) is considerable. In the United States this constant TV-related leaking of *electricity* requires about two gigawatts, and the worldwide burden of such phantom loads is close to six gigawatts, an equivalent of all installed generating capacity in Vietnam or Nigeria.

Computers and Microchips

The term *computer* is now a functional misnomer, inasmuch as most of today's *computers* are not used by their owners to manipulate numbers. But this mass diffusion of nonmathematical applications is a very recent phenome-

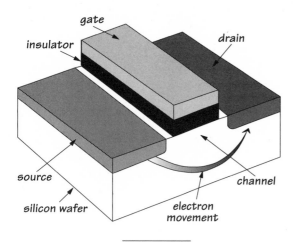

The field-effect transistor is a sandwich of variously doped layers of silicon.

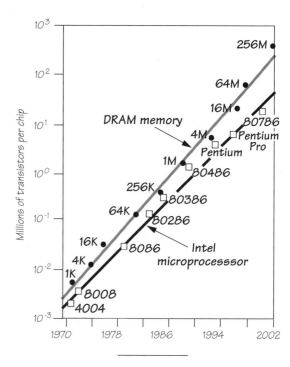

Moore's law: Between 1970 and 1995 the density of transistor circuits on a single chip rose more than ten thousand times.

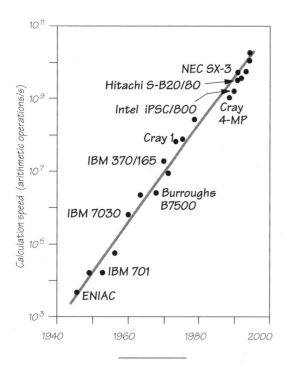

Since 1945 the fastest computer speeds have increased by more than seven orders of magnitude.

non compared with the long history of machines designed to speed up complex calculations. After 1820 it appeared that new ingenious mechanical calculators will finally become a commercial reality—but neither the abandoned prescient designs of difference engines by Charles Babbage nor the actually built and operated calculating machines of Georg and Edward Schuetz served as models of further advances.

The quest for rapid calculation, progressing slowly since the end of the nineteenth century, was relaunched in earnest only during World War II with the construction of the first electronic *computers*. The calculating speed of these programmable machines began to rise exponentially only after transistors supplanted vacuum tubes, and personal *computers* became both powerful and widely affordable only with the progressive miniaturization of transistors, resistors, and capacitors and with a rapid development of integrated circuits placed on a *microchip* of silicon. Starting with a single planar transistor in 1959, the number of components per *microchip* doubled every year until 1972.

Since that year it has continued to increase at only a slightly lower rate, doubling every eighteen months. This rate, predicted by Intel's co-founder, became known as Moore's law. By the early 1980s more than 100,000 components were fitted onto a single DRAM memory *microchip,* by 1990 the total reached one million, and by the year 2000 it will top 100 million. During the same period the number of transistors per Intel's microprocessor chip will have risen from fewer than fifty thousand to ten million.

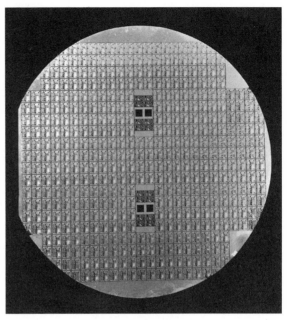

Silicon wafer with 470 microchips.

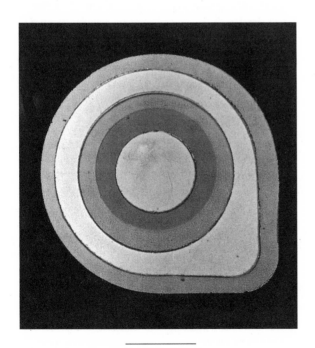

The first planar transistor, made by Fairchild Semiconductor Corporation in 1959.

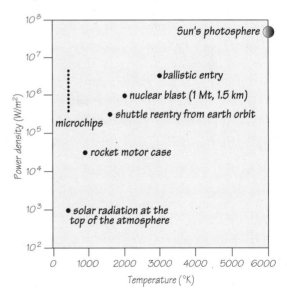

Power densities of heat rejection.

These trends have brought rapidly declining costs creating huge new markets for *microchips* beyond personal *computers. Microchips* have found new applications in consumer electronics, telecommunication, precise manufacturing, and process controls. This computing revolution started inconspicuously on January 27, 1947, when three Bell Labs physicists revealed the first transistor, a device transferring electric signals across a resistor. Even the first prototype occupied only one-two hundredths of the space of an equally effective vacuum tube while its solid state assured higher reliability (deterioration of hot cathodes tubes led to lives of just fifty to a hundred hours) and lower energy use without any warmup time.

When extremely pure, germanium and silicon are insulators; when contaminated ("doped") with minuscule quantities of other atoms (arsenic, boron, phosphorus) they conduct. Physical properties make silicon a particularly good choice for making transistors, the simple switches at the heart of all modern electronics. They begin by growing high-purity tubular silicon crystals, up to one meter long and fifteen centimeters in diameter. These crystals are sliced into microthin wafers that are etched with a steadily rising number of integrated circuits. If Moore's law is to remain in effect designers will have to substitute the optical lithography used to pattern silicon wafers with features as narrow as 0.25 micrometers by even more discriminating techniques.

Intricate *microchip* design and sophisticated production has emerged as perhaps the most emblematic industry of the late twentieth century. This material revolution is far from exhausted. Silicon-germanium alloys are materials that allow sharply increased velocities of electrons and can be used to make exceptionally high-speed transistors with existing equipment.

Naturally, reliable *electricity* supply is the critical precondition for using all *microchips*—but their global profusion puts little strain on *electricity* demand. Power ratings of microcomputers, typically just 10^2 watts, are very low in comparison with both their predecessors and with other common electrical gadgets. And microchip-driven controls are a major ingredient in the continuing quest for more efficient use of energy. Consequently, it is not *electricity* supply, but rather *heat disposal,* that is a challenge for designers.

The most concentrated units among these tiny packages, very large scale integration circuits, have to dissipate heat at rates of about one megawatt per square meter. This is a thermal loading higher than the space shuttle's shielding experiences when reentering the atmosphere, and the rate is just an order of magnitude lower than the flux through the *Sun*'s photosphere. A *microchip* tethered in still air and cooled just by natural convection would be thus obliterated by being heated to more than 6000°C. When *microchips* are installed in a *computer* their temperature is kept no higher than 100°C by a combination of forced convection using circuitous airflow (hence the constant hum coming from the CPU box) and heat conduction through tiny spring-loaded pistons or solder bumps taking heat away to cold plates.

FURTHER READING

In the following lists I have combined the latest surveys and analyses with references to many key books that appeared during the 1970s and 1980s, the two golden decades of modern energy studies. A few older classics are also included, together with a small number of particularly interesting articles.

Energies

History of Energy Use

Basalla, G. 1988. *The Evolution of Technology*. Cambridge: Cambridge University Press.

Derry, T. K. and T. I. Williams. 1960. *A Short History of Technology*. Oxford: Oxford University Press.

Finniston, M., et al., eds. 1992. *Oxford Illustrated Encyclopedia of Invention and Technology*. Oxford: Oxford University Press.

Forbes, R. J. 1964–1972. *Studies in Ancient Technology*. Leiden: E. J. Brill.

Goudsblom, J. 1992. *Fire and Civilization*. London: Allen Lane.

Jones, H. M. 1970. *The Age of Energy*. New York: Viking.

Needham, J. 1954– . *Science and Civilization in China*. Cambridge: Cambridge University Press.

Pacey, A. 1990. *Technology in World Civilization*. Cambridge, Mass.: MIT Press.

Singer, C., et al., eds. 1954–1984. *A History of Technology*. Oxford: Clarendon Press.

Smil, V. 1993. *Energy in World History*. Boulder: Westview.

Williams, T. I. 1987. *The History of Invention: From Stone Axes to Silicon Chips*. New York: Facts on File.

Energy and Society

Cook, E. 1976. *Man, Energy, Society*. San Francisco: W. H. Freeman.

Cottrell, F. 1955. *Energy and Society*. New York: McGraw-Hill.

Georgescu-Roegen, N. 1971. *The Entropy and the Economic Process*. Cambridge, Mass.: Harvard University Press.

Lovins, A. B. 1977. *Soft Energy Paths: Toward a Durable Peace*. Cambridge, Mass.: Friends of the Earth and Ballinger.

McLaren, D. J., and B. J. Skinner, eds. 1987. *Resources and World Development*. Chichester: John Wiley.

Rose, D. J. 1986. *Learning about Energy*. New York: Plenum.

Smil, V. 1987. *Energy Food Environment: Realities Myths Options*. Oxford: Oxford University Press.

Thirring, H. 1958. *Energy for Man*. Bloomington: Indiana University Press.

Ubbelohde, A. R. 1954. *Man and Energy.* London: Hutchinson.

World Energy Council Commission. 1993. *Energy for Tomorrow's World.* London: Kogan Page.

Energy Analysis
Boustead, I., and G. F. Hancock. 1979. *Handbook of Industrial Energy Analysis.* Chichester: Ellis Horwood.

Brooks, D. R., and E. O. Wiley. 1986. *Evolution as Entropy.* Chicago: University of Chicago Press.

Feynman, R. 1988. *The Feynman Lectures on Physics.* Reading, Mass.: Addison-Wesley.

Krenz, J. H. 1976. *Energy Conversion and Utilization.* Boston: Allyn and Bacon.

Lotka, A. 1925. *Elements of Physical Biology.* Baltimore: Williams and Wilkins.

Odum, H. T. 1971. *Environment, Power, and Society.* New York: John Wiley.

Pimentel, D., ed. 1980. *Handbook of Energy Utilization in Agriculture.* Boca Raton, Fla.: CRC Press.

Smil, V. 1991. *General Energetics.* New York: John Wiley.

Thomas, J., ed. 1979. *Energy Analysis.* Boulder: Westview.

Wesley, J. P. 1974. *Ecophysics.* Springfield, Ill.: Charles C. Thomas.

Sun and Earth
Sun and Earth
Emiliani, C. 1992. *Planet Earth: Cosmology, Geology, and the Evolution of Life and Environment.* New York: Cambridge University Press.

Lewis, J. S. 1997. *Rain of Iron and Ice: The Very Real Threat of Comet and Asteroid Bombardment.* Reading, Mass.: Addison-Wesley.

IAU Colloquium. 1994. *The Sun as a Variable Star.* New York: Cambridge University Press.

Tayler, R. J. 1997. *The Sun as a Star.* New York: Cambridge University Press.

Transformations of Solar Energy
Baker, V. R., ed. 1981. *Catastrophic Flooding.* Stroudsburg, Pa.: Dowden, Hutchison & Ross.

Barton, R. 1980. *The Oceans.* New York: Facts on File.

Berner, E. K., and R. A. Berner. 1987. *The Global Water Cycle.* Englewood Cliffs, N.J.: Prentice Hall.

Caine, N. 1976. A uniform measure of subaerial erosion. *Geological Society of America Bulletin* 87: 137–40.

Czaya, E. 1981. *Rivers of the World.* New York: Van Nostrand Reinhold.

Das, P. K. 1995. *The Monsoons.* New Delhi: National Book Trust.

Denny, M. W. 1993. *Air and Water: The Biology and Physics of Life's Media.* Princeton, N.J.: Princeton University Press.

Desbois, M., and F. Desalmand, eds. 1994. *Global Precipitations and Climate Change.* Berlin: Springer-Verlag.

Green, J. 1998. *Atmospheric Dynamics.* New York: Cambridge University Press.

Grotjahn, R. 1993. *Global Atmospheric Circulations: Observations and Theories.* Oxford University Press, New York.

Hickin, E. J., ed. 1995. *River Geomorphology.* Chichester: John Wiley.

Kessler, A. 1985. *Heat Balance Climatology.* Amsterdam: Elsevier.

Longshore, D. 1998. *Encyclopedia of Hurricanes, Typhoons and Cyclones.* New York: Facts on File.

Lorenz, E. N. 1976. *The Nature and Theory of the General Circulation of the Atmosphere.* Geneva: WMO.

Miller, D. H. 1977. *Water at the Surface of the Earth.* New York: Academic Press.

———. 1981. *Energy at the Surface of the Earth.* New York: Academic Press.

Ross, S. 1989. *Soil Processes: A Systematic Approach.* London: Routledge.

Selby, M. J. 1985. *Earth's Changing Surface.* Oxford: Clarendon Press.

Smil, V. 1997. *Cycles of Life.* New York: Scientific American Library.

U.S. National Environmental Satellite, Data, and Information Service. 1994. *World Ocean Atlas 1994.* Washington, D.C.: U.S. Dept. of Commerce.

The Earth's Heat and Geotectonics

Decker, R. W. 1998. *Volcanoes.* New York: W. H. Freeman.

Condie, K. C. 1997. *Plate Tectonics and Crustal Evolution.* Oxford: Butterworth Heinemann.

Francis, P. 1993. *Volcanoes: A Planetary Perspective.* Oxford: Clarendon Press.

Kanamori, H., and E. Boschi, eds. 1983. *Earthquakes: Observation, Theory and Interpretation.* Amsterdam: North-Holland.

Olsen, K. H., ed. 1995. *Continental Rifts: Evolution, Structure, Tectonics.* Amsterdam: Elsevier.

Richter, C. F. 1935. An instrumental earthquake magnitude scale. *Bulletin Seismological Society of America* 25: 1–32.

Scheidegger, A. E. 1982. *Principles of Geodynamics.* Berlin: Springer-Verlag.

Verhoogen, J. 1980. *Energetics of the Earth.* Washington, D.C.: National Academy of Sciences.

Yeats, R. S., et al. 1997. *The Geology of Earthquakes.* New York: Oxford University Press.

Plants and Animals

Photosynthesis, Plants, and Bacteria

Baker, N. R., ed. 1996. *Photosynthesis and the Environment.* Dordrecht: Kluwer Academic Publishers.

Bassham, J. A. and M. Calvin. 1957. *The Path of Carbon in Photosynthesis.* Englewood Cliffs, N.J.: Prentice-Hall.

Coupland, R. T., ed. 1980. *Ecosystems of the World: Analysis of Grasslands and their Uses.* Cambridge: Cambridge University Press.

———. 1992–1993. *Natural Grasslands.* Amsterdam: Elsevier.

de Duve, C. 1984. *A Guided Tour of the Living Cell.* New York: Scientific American Library.

Dixon, B. 1994. *Power Unseen: How Microbes Rule the World.* New York: W. H. Freeman.

Edwards, G., and D. Walker. 1983. C_3, C_4: *Mechanisms and Cellular and Environmental Regulation of Photosynthesis.* Berkeley: University of California Press.

Falkowski, P. G. and J. A. Raven. 1997. *Aquatic Photosynthesis.* Walden, Mass.: Blackwell Science.

Fogg, G. E. 1987. *Algal Cultures and Phytoplankton Ecology.* Madison: University of Wisconsin Press.

Gates, D. M. 1985. *Energy and Ecology.* Sunderland, Mass.: Sinauer Associates.

Harris, G. P. 1986. *Phytoplankton Ecology: Structure, Function, and Fluctuation.* London: Chapman and Hall.

Landsberg, J. J. 1986. *Physiological Ecology of Forest Production.* New York: Academic Press.

Persson, T., ed. 1980. *Structure and Function of Northern Coniferous Forests.* Stockholm: Ecological Bulletins.

Pessarakli, M., ed. 1997. *Handbook of Photosynthesis.* New York: Marcel Dekker.

Raghavendra, A. S., ed. 1997. *Photosynthesis: A Comprehensive Treatise.* New York: Cambridge University Press.

Reichle, D. E., ed. 1981. *Dynamic Properties of Forest Ecosystems.* Cambridge: Cambridge University Press.

Animal Energetics

Andrews, J. H. 1991. *Comparative Ecology of Microorganisms and Macroorganisms.* New York: Springer-Verlag.

Brafield, A. E. and M. J. Llewellyn. 1982. *Animal Energetics.* Glasgow: Blackie.

Broda, E. 1978. *The Evolution of Bioenergetic Processes.* Oxford: Pergamon Press.

Brody, S. 1945. *Bioenergetics and Growth.* New York: Reinhold, New York.

Campion, D. R. et al., eds. 1989. *Animal Growth Regulation.* New York: Plenum.

Cooke, J. A. L. 1991. *The Restless Kingdom: An Exploration of Animal Movement.* New York: Facts on File.

Crawley, M. J. 1983. *Herbivory: The Dynamics of Animal-Plant Interactions.* Berkeley: University of California Press.

Dettner. K., et al., eds. 1997. *Vertical Food Web Interactions: Evolutionary Patterns and Driving Forces.* Berlin: Springer-Verlag.

Dusenbery, D. B. 1996. *Life at Small Scale.* New York: Scientific American Library.

Hacker, J. B. and J. H. Ternouth, eds. 1987. *The Nutrition of Herbivores.* Orlando: Academic Press.

Hudson, R. J., and R. G. White, eds. 1985. *Bioenergetics of Wild Herbivores*. Boca Raton, Fla.: CRC Press.

Kleiber, M. 1961. *The Fire of Life*. New York: John Wiley.

Levitt, J. R. 1989. *Carnivores*. New York: St. Martin's Press.

McMahon, T., and J. T. Bonner. 1983. *On Size and Life*. New York: Scientific American Library.

McNeill, A. R. 1992. *Exploring Biomechanics: Animals in Motion*. New York: Scientific American Library.

Owen-Smith, R. N. 1988. *Megaherbivores: The Influence of Very Large Body Size on Ecology*. New York: Cambridge University Press.

Reagan, D. P., and R. B. Waide, eds. 1996. *The Food Web of a Tropical Rain Forest*. Chicago: University of Chicago Press.

Robbins, C. T. 1983. *Wildlife Feeding and Nutrition*. New York: Academic Press.

Ruppell, G. 1977. *Bird Flight*. New York: Van Nostrand Reinhold.

Schmidt-Nielsen, K. 1984. *Scaling: Why is Animal Size so Important?* Cambridge: Cambridge University Press.

Videler, J. J. 1993. *Fish Swimming*. New York: Chapman & Hall.

West, G. B., et al. 1997. A general model for the origin of allometric scaling laws in biology. *Science* 276: 122–126.

People and Food
Human Energetics

Alexander, R. M. 1992. The *Human Machine*. New York: Columbia University Press.

Durnin, J. V. G. A., and R. Passmore. 1967. *Energy, Work and Leisure*. London: Heinemann Educational Books.

Carrier, D. R. 1984. The energetic paradox of human running and hominid evolution. *Current Anthropology* 25: 483–95.

Frigerio, C. et al. 1991. Is human lactation a particularly efficient process? *European Journal of Clinical Nutrition* 45: 459–62.

Glover, B., et al. 1996. *The Runner's Handbook*. New York: Penguin Books.

Hanna, J. M. and D. E. Brown. 1983. Human heat tolerance: an anthropological perspective. *Annual Review of Anthropology* 12: 259–84.

Hardy, J. D. et al. 1971. Man. In G. C. Whittow, ed., *Comparative Physiology of Thermoregulation,* pp. 327–80. New York: Academic Press.

Hytten, F. E. and I. Leitch. 1964. *The Physiology of Human Pregnancy*. Edinburgh: Blackwell.

Joint FAO/WHO/UNU Expert Consultation. 1985. *Energy and Protein Requirements*. Geneva: WHO.

Neville, M. C., and M. R. Neifert, eds. 1983. *Lactation*. New York: Plenum.

Pollitt, E., and P. Amanate, eds. 1984. *Energy Intake and Activity*. New York: Alan R. Liss.

Schofield, W. N., et al. 1985. Basal metabolic rate in man: survey of the literature and computation of equations for prediction. *Human Nutrition Clinical Nutrition* 39 (Supplement 1): 1–96.

Schroeder, B. A. 1992. *Human Growth and Development*. St. Paul: West Publishing.

Ulijaszek, S., et al., eds. 1997. *The Cambridge Encyclopedia of Human Growth and Development*. New York: Cambridge University Press.

Foodstuffs and Nutrition

Bourne, G. H., ed. 1989. *Nutritional Value of Cereal Products, Beans, and Starches*. Basel: Karger.

Food and Agriculture Organization. 1950– . *Production Yearbook*. Rome: FAO.

———. 1996. *Food Requirements and Population Growth*. Rome: FAO.

Hensperger, B. 1992. *Baking Bread: Old and New Traditions*. San Francisco: Chronicle Books.

Hui, Y. H., ed. 1993. *Dairy Science and Technology Handbook*. New York: VCH.

Litten, R. Z., and J. P. Allen, eds. 1992. *Measuring Alcohol Consumption: Psychosocial and Biochemical Methods*. Totowa, N.J.: Humana Press.

Matthews, R. H., ed. 1989. *Legumes*. New York: Marcel Dekker.

National Research Council. 1989. *Recommended Dietary Allowances*. Washington, D.C.: National Research Council.

Rosenthal, I. 1991. *Milk and Dairy Products: Properties and Processing.* Weinheim: VCH.

Snyder, H. 1930. *Bread.* New York: Macmillan.

Vance, D. E., and J. E. Vance, eds. 1996. *Biochemistry of Lipids, Lipoproteins, and Membranes.* Amsterdam: Elsevier.

Watt, B. K. and A. L. Merrill. 1975. *Handbook of the Nutritional Contents of Foods.* New York: Dover Publications.

Preindustrial Societies
Food and Animate Energies

Bettinger, R. L. 1991. *Hunter-Gatherers Archaeological and Evolutionary Theory.* New York: Plenum.

Bray, F. 1984. *Science and Civilisation in China.* Volume 6, Part II: *Agriculture.* Cambridge: Cambridge University Press.

Clark, J. D., and S. A. Brandt, eds. 1984. *From Hunters to Farmers.* Berkeley: University of California Press.

Cowan, C. W., and P. J. Watson, eds. 1992. *The Origins of Agriculture.* Washington, D.C.: Smithsonian Institution Press.

Dent, A. 1974. *The Horse.* New York: Holt, Rinehart and Winston.

Grigg, D. B. 1974. *The Agricultural Systems of the World.* Cambridge: Cambridge University Press.

Headland, T. N., and L. A. Reid. 1989. Hunter-gatherers and their neighbors from prehistory to the present. *Current Anthropology* 30: 43–66.

Langdon, J. 1986. *Horses, Oxen and Technological Innovation.* Cambridge: Cambridge University Press.

Price, T. D., and J. A. Brown, eds. 1985. *Prehistoric Hunter-Gatherers.* Orlando: Academic Press.

Rouse, J. E. 1970. *World Cattle.* Norman: University of Oklahoma Press.

Silver, C. 1976. *Guide to the Horses of the World.* Oxford: Elsevier Phaidon.

Telleen, M. 1977. *The Draft Horse Primer.* Emmaus, Pa.: Rodale Press.

Watters, R. F. 1971. *Shifting Cultivation in Latin America.* Rome: FAO.

Inanimate Energies and Materials

Barjot, A. and J. Savant. 1965. *History of the World's Shipping.* London: Ian Allan.

Chatterton, E. K. 1977. *Sailing Ships: The Story of their Development from the Earliest Times to the Present.* New York: Gordon Press.

Cipolla, C. M. 1966. *Guns, Sails and Empires.* New York: Pantheon Books.

Forbes, R. J. 1972. Copper. In R. J. Forbes, *Studies in Ancient Technology,* vol. 9, pp. 1–133. Leiden: E. J. Brill.

Gustavson, M. R. 1979. Limits to wind power utilization. *Science* 204: 13–17.

Hindle, B., ed. 1975. *America's Wooden Age: Aspects of Its Early Technology.* Tarrytown, N.Y.: Sleepy Hollow Restorations.

McNeill, W. H. 1989. *The Age of Gunpowder Empires 1450–1800.* Washington, D.C.: American Historical Association.

Reynolds, J. 1970. *Windmills and Watermills.* London: Hugh Evelyn.

Schubert, H. R. 1958. Extraction and production of metals: iron and steel. In C. Singer et al., eds., *A History of Technology,* vol. IV, pp. 99–117, vol. V, pp. 53–71. Oxford: Oxford University Press.

Skilton, C. P. 1947. *British Windmills and Watermills.* London: Collins.

Tillman, D. A. 1978. *Wood as an Energy Resource.* New York: Academic Press

Torrey, V. 1976. *Wind-Catchers. American Windmills of Yesterday and Tomorrow.* Brattleboro, Vt.: Stephen Greene Press.

Wilson, P. N. 1956. *Watermills: An Introduction.* Mexborough, Pa.: Times Printing.

Fossil-Fueled Civilization
Primary Energies

Brantly, J. E. 1971. *History of Oil Well Drilling.* Houston: Gulf Publishing.

Fettweis, G. B. 1979. *World Coal Resources.* Amsterdam: Elsevier.

Gold, T. 1987. *Power from the Earth: Deep Earth Gas — Energy for the Future.* London: J. M. Dent and Sons.

Gordon, R. L. 1987. *World Coal: Economics, Policies, and Prospects.* New York: Cambridge University Press.

Hubbert, M. K. 1962. *Energy Resources; A Report to the Committee on Natural Resources.* Washington, D.C.: National Academy of Sciences.

McLaren, D. J., and B. J. Skinner, eds. 1987. *Resources and World Development.* Chichester: John Wiley.

Nehring, R. 1978. *Giant Oil Fields and World Oil Resources.* Santa Monica: Rand Corporation.

Perrodon, A. 1985. *Histoire des Grandes Découvertes Petrolieres.* Paris: Elf Aquitaine.

Riva, J. P. 1983. *World Petroleum Resources and Reserves.* Boulder: Westview.

Ward, C. R., ed. 1984. *Coal Geology and Coal Technology.* Melbourne: Blackwell Scientific.

World Energy Council. 1992. *Survey of Energy Resources.* London: World Energy Council.

Energy Conversions

Berkowitz, N. 1997. *Fossil Hydrocarbons: Chemistry and Technology.* San Diego: Academic Press.

Cohen, H. et al. 1987. *Gas Turbine Theory.* New York: Longman.

Devine, W. D. 1983. From shafts to wires: historical perspective on electrification. *The Journal of Economic History* 63: 347–72.

Ganesan, V. 1996. *Internal Combustion Engines.* New York: McGraw-Hill.

Gunston, B. 1986. *World Encyclopaedia of Aero Engines.* Wellingborough, England: Patrick Stephens.

Jones, H. 1973. *Steam Engines.* London: Ernest Benn.

Josephson, M. 1959. *Edison: A Biography.* New York: McGraw-Hill.

King, B. R., ed. 1993. *The Steam Turbine-Generator Today.* New York: American Society of Mechanical Engineers.

National Research Council. 1986. *Electricity in Economic Growth.* Washington, D.C.: National Academy Press.

Nye, D. E. 1992. *Electrifying America.* Cambridge, Mass.: MIT Press.

Pansini, A. J. 1996. *Basics of Electric Motors.* Tulsa: Pennwell.

Pansini, A. J., and K. G. Samlling. 1994. *High Voltage Power Equipment Engineering.* Englewood Cliffs, N.J.: Fairmont Press.

Pulkrabek, W. W. 1997. *Engineering Fundamentals of the Internal Combustion Engine.* Upper Saddle River, N.J.: Prentice Hall.

Robinson, E. and A. E. Musson. 1969. *James Watt and the Steam Revolution.* New York: Augustus M. Kelley.

Schipper, L., and S. Meyers. 1992. *Energy Efficiency and Human Activity.* New York: Cambridge University Press.

Smith, N. 1980. The origins of the water turbine. *Scientific American* 242 (1): 138–48.

Watkins, G. 1967. Steam power — an illustrated guide. *Industrial Archaeology* 4 (2): 81–110.

Wharton, A. J. 1991. *Diesel Engines.* Oxford: B. H. Newnes.

Wymer, R. G., and B. L. Vondra Jr., eds. 1981. *Light Water Reactor Nuclear Fuel Cycle.* Boca Raton, Fla.: CRC Press.

Materials and Weapons

Atkins, P. R., et al. 1991. Some energy and environmental impacts of aluminum usage. In J. W. Tester et al., eds. *Energy and the Environment in the 21st Century,* pp. 383–87. Cambridge, Mass.: MIT Press.

Committee for the Compilation of Materials on Damage Caused by the Atomic Bombs in Hiroshima and Nagasaki. 1981. *Hiroshima and Nagasaki.* New York: Basic Books.

Food and Agriculture Organization. 1970–. *Fertilizer Yearbook.* Rome: FAO.

Heal, D. W. 1975. Modern perspectives on the history of fuel economy in the iron and steel industry. *Ironmaking and Steelmaking* 2: 222–27.

Helsel, Z., ed. 1987. *Energy in Plant Nutrition and Pest Control.* Amsterdam: Elsevier.

Kesaris, P., ed. 1977. *Manhattan Project: Official History and Documents.* Washington, D.C.: University Publications of America.

National Academy of Sciences. 1980. *Materials Aspects of World Energy Needs.* Washington, D.C.: National Academy Press.

Nitsche, J. 1992. *Production of Aluminium*. Düsseldorf: Aluminium-Verlag.

Peacey, J. G. 1979. *The Iron Blast Furnace*. Oxford: Pergamon Press.

Strategic Air Command. 1990. *SAC Missile Chronology, 1939–1988*. Offutt Air Force Base, Neb.: Strategic Air Command.

United Nations Organization. 1980– . *Yearbook of World Energy Statistics*. New York: UNO.

Transportation and Information
Transportation
Baker, D. 1996. *Spaceflight and Rocketry: A Chronology*. New York: Facts on File.

Cairis, N. T. 1992. *Era of the Passenger Liner*. London: Pegasus Books.

Constant, E. W. 1981. *The Origins of the Turbojet Revolution*. Baltimore: Johns Hopkins University Press.

Coulbeck, B., and E. Evans, eds. 1992. *Pipeline Systems*. Dordrecht: Kluwer Academic.

Drela, M. and J. S. Langford. 1985. Human-powered flight. *Scientific American* 253 (5): 144–51.

Ellis, C. H. 1977. *The Lore of the Train*. New York: Crescent Books.

Flink, J. J. 1988. *The Automobile Age*. Cambridge, Mass.: MIT Press.

Flower, R. and M. W. Jones. 1981. *One Hundred Years of Motoring*. Maidenhead, England: McGraw-Hill.

Fry, H. 1896. *The History of North Atlantic Steam Navigation*. London: Sampson, Low, Marston & Co.

Haney, L. H. 1968. *A Congressional History of Railways in the United States*. New York: A. M. Kelley.

Kienow, K. K., ed. 1990. *Pipeline Design and Installation*. New York: American Society of Civil Engineers.

Motor Vehicle Manufacturing Association. 1970– . *Facts & Figures '97*. Detroit: MVMA.

Ratcliffe, K. 1985. *Liquid Gold Ships: History of the Tanker (1859–1984)*. London: Lloyds.

Sullivan, G. 1978. *Supertanker: The Story of the World's Biggest Ships*. New York: Dodd, Mead.

Tennekes, H. 1996. *The Simple Science of Flight*. Cambridge, Mass.: MIT Press.

von Braun, W., and F. I. Ordway. 1975. *History of Rocketry and Space Travel*. New York: Thomas Y. Crowell.

Whitt, F. R., and D. G. Wilson. 1982. *Bicycling Science*. Cambridge, Mass.: MIT Press.

Wright, O. 1953. *How We Invented the Airplane*. New York: David McKay.

Communication and Computing
Collins, P. 1997. *Radios: The Golden Age*. New York: Black Dog & Leventhal Publishers.

Dooner, K. E. 1992. *Telephones: Antique to Modern*. West Chester, Pa.: West Chester.

Freed, L. 1995. *The History of Computers*. Emeryville, Calif.: Ziff-Davis Press.

Hounshell, D. A. 1981. Two paths to the telephone. *Scientific American* 244 (1): 156–64.

Locke, I. 1995. *The Microchip and How it Changed the World*. New York: Facts on File.

Oktay, S., et al. 1986. High heat from a small package. *Mechanical Engineering* 108 (3): 36–42.

Van Zant, P. 1997. *Microchip Fabrication: A Practical Guide to Semiconductor Processing*. New York: McGraw-Hill.

Index